Get the eBook FREE!
(PDF, ePub, Kindle, and liveBook all included)

We believe that once you buy a book from us, you should be able to read it in any format we have available. To get electronic versions of this book at no additional cost to you, purchase and then register this book at the Manning website.

Go to https://www.manning.com/freebook and follow the instructions to complete your pBook registration.

That's it!
Thanks from Manning!

Cybersecurity Career Guide

Cybersecurity Career Guide

ALYSSA MILLER

MANNING

SHELTER ISLAND

Manning Publications Co.   Development editor: Karen Miller
20 Baldwin Road   Review editor: Aleksandar Dragosavljević
PO Box 761   Production editor: Deirdre S. Hiam
Shelter Island, NY 11964   Copy editor: Sharon Wilkey
                                           Proofreader: Jason Everett
                                            Typesetter: Dennis Dalinnik
                                        Cover designer: Marija Tudor

ISBN: 9781617298202
Printed in the United States of America

brief contents

contents

dedication

The many skills and experiences I've gained over the years that made this guide possible were due in no small part to the inspiration of three wonderful leaders I have worked with in my career. I dedicate this book to them.

To Tim Patneaude, who took a risk in hiring a young, hungry programmer with no degree into his home banking product development team. Tim, you taught me my earliest lessons of how to be an empowering servant leader, and that has immeasurably shaped the course of my career.

To Donna Beaumeister, who nine years later took a chance on a programmer with no security experience and put me at the lead of her security test team. Your empathetic, transparent, yet confident leadership style taught me how to be a credible leader and allow my teams to be authentic and powerful forces for progress.

And finally, to Patrick Fleming, you showed me everything that an inspiring leader should be and how to bring my own vulnerability in a way that would drive my teams to new heights.

Thank you, all three, for all that you've meant in helping me grow professionally and personally into the person I am today. I will never forget the gifts each of you has given me.

preface

I grew up as a hacker. I have been working with computers all my life and have been a cybersecurity professional for the better part of two decades. My time spent in the hacker communities of the 1990s along with my own successful career progression have etched a special place in my heart for the cybersecurity field.

I've watched as cybersecurity has gone from a relatively unknown field, where I often struggled to even explain what it was I did for a living, to now an almost ubiquitous presence in the mainstream lives of every person on the planet. It is no secret that cybersecurity has become one of the most talked about career fields in technology and maybe just in general society.

However, as technology continues to grow, innovate, and permeate our daily lives in so many ways, the need for cybersecurity professionals has skyrocketed. I have read many accounts in industry and mainstream media of this so-called skills gap that we have in cybersecurity. I've seen estimates of anywhere from hundreds of thousands to millions of unfilled jobs. Yet, as I spoke with people who were trying to get into this hot job market, I heard stories of the struggles they were encountering. A sort of dissonance exists between what the industry is saying and what job seekers are experiencing. So I set out to solve this issue.

Through surveys, interviews, and other research, I've sought the answers. My goal is to help connect job seekers and the companies that need to hire them. One of the primary efforts of my ongoing work to improve the cybersecurity community is this guide. The *Cybersecurity Career Guide* is here to help those who want to start their career understand, identify, and overcome the many obstacles and challenges they will face.

My hope in writing this book is to ease that path and guide those new professionals who will help secure our digital world.

acknowledgments

A book like this does not come together as the result of any one person's efforts. Thus, I take a moment here to thank all the amazing people who made these pages possible.

I begin with Casey Shuniak, whose love, support, and guiding ideas helped make this book so much better than it might have been otherwise. Casey's support in helping me navigate many moments during the COVID-19 pandemic in which I lost motivation ensured that this book got out to the many who will benefit from it. Her efforts in reviewing the content of this book provided me with invaluable perspectives on how to connect with a broader audience of readers. For that, I am most deeply thankful.

I also want to acknowledge the many people in the cybersecurity industry who helped in various ways through the interviews I conducted and the research I performed. In no particular order, Alethe Denis, Carolina Terrazas, Kirsten Renner, Kwadwo Burgee, Kathleen Smith, Gabrielle Hempel, Keenan Skelly, Mitch Parker, and Lesley Carhart, thank you for all the knowledge and perspectives you shared that can be found throughout the pages of this guide. Additionally, special thanks to Carl Hertz and Ray [REDACTED] for being tremendous advocates of my work, sounding boards when I needed an outside opinion, and simply two of the most wonderful people I could ever hope to have in my court cheering me on.

I would also like to thank the people at Manning who helped shepherd me through this process: Karen Miller, the development editor; Aleksandar Dragosavljević, the review editor; Deirdre Hiam, my project manager; Sharon Wilkey, my copyeditor; and Jason Everett, my proofreader.

It is also important for me to thank the thousands of people who participated in the research surveys that really kickstarted this book and provided some surprising data and conclusions. So many revelations came from your efforts in sharing your experiences, and I can't thank you enough for helping me pay it forward to help others navigate their way into a cybersecurity career. And then, finally, thank you to the tens of thousands of members of the security community I am blessed to be able to interact with on a daily basis through social media, professional events, and industry media. This community I love so dearly served as the true inspiration for writing this guide, and I truly appreciate every single one of you.

To all the reviewers: Alex Saez, Amanda Debler, Amit Lamba, Björn Neuhaus, Bobby Lin, Daniel Varga, Dhivya Sivasubramanian, Dipen N. Kumar, Elavarasu A K, Emanuele Origgi, Georgerobert Freeman, Harsh Raval, Hugo Sousa, Jan Vinterberg, Jason Hales, Joerg Discher, Marc Roulleau, Michele Di Pede, Nathan Delboux, Paul Ammann, Rafik Naccache, Rob Goelz, Steve Atchue, and Zarak Mahmud, your suggestions helped make this a better book.

about this book

Cybersecurity Career Guide will assist you in uncovering your path to becoming a great security practitioner. You'll learn how to reliably enter the security field and quickly grow into your new career, following clear, practical advice that's based on independent research and interviews with hundreds of hiring managers.

Who should read this book

The book is written as a complete guide to help any individual who is looking to launch a career in cybersecurity. Individuals from all walks of life, all technical skill levels, and all knowledge levels will find something in this book that can help start their career in cybersecurity. Whether this is your first career path or you are trying to pivot from another industry, this guide is written to ensure that you are equipped with the tools you need to succeed in your first job search and to position yourself for long-term growth.

How this book is organized

This book is divided into three parts of three chapters each. The first part provides a common understanding of cybersecurity. You will learn just how expansive this domain of knowledge is and see the details of many varied paths you can pursue when embarking on a cybersecurity career. This part also discusses many of the challenges and obstacles you may encounter along your journey. Research is presented to illustrate some of the difficulties that new entrants face.

Part 2 focuses on you as the job seeker. In the chapters of this part, you will find practical exercises that you can use to discover your passions and interests and map those to the various disciplines of cybersecurity. The goal is to focus your efforts and maximize the chances that you pick a discipline that will best align with your capabilities. You will also learn how to structure your job search, create your resume, and prepare for and ace the interview process.

In the final three chapters that make up part 3, you will learn how to set yourself up to achieve the goals of a long-term career in cybersecurity. This part presents tools for understanding and leveraging networking and mentoring to drive your journey. You will also be given powerful tools for overcoming one of the most damaging influences that derails careers, impostor syndrome. Then, in the final chapter of this guide, everything comes together as you'll learn to set realistic goals and ensure that your journey stays on a path to new and greater accomplishments.

liveBook discussion forum

Purchase of *Cybersecurity Career Guide* includes free access to liveBook, Manning's online reading platform. Using liveBook's exclusive discussion features, you can attach comments to the book globally or to specific sections or paragraphs. It's a snap to make notes for yourself, ask and answer technical questions, and receive help from the author and other users. To access the forum, go to https://livebook.manning.com/book/cyber-security-career-guide/discussion/. You can also learn more about Manning's forums and the rules of conduct at https://livebook.manning.com/discussion.

Manning's commitment to our readers is to provide a venue where a meaningful dialogue between individual readers and between readers and the author can take place. It is not a commitment to any specific amount of participation on the part of the author, whose contribution to the forum remains voluntary (and unpaid). We suggest you try asking the author some challenging questions lest her interest stray! The forum and the archives of previous discussions will be accessible from the publisher's website as long as the book is in print.

Other online resources

Several online resources are available to help you embark on your cybersecurity career. The following are three that can help get you started:

- *Cyber Career Pathways Tool*—This tool from the National Initiative for Cybersecurity Careers and Studies (NICCS) and hosted by the US Cybersecurity and Infrastructure Security Agency (CISA) can help you explore and map out possible career journeys (https://niccs.cisa.gov/workforce-development/cyber-career-pathways).
- *TryHackMe*—This free online training website is designed to help cybersecurity professionals learn various skills through short lessons and challenges (https://tryhackme.com).

- *CompTIA Resource Center*—You can use this collection of free educational resources from the Computing Technology Industry Association (CompTIA) to learn more about cybersecurity and cybersecurity-adjacent technologies and fields (www.comptia.org/resources).

about the author

A hacker since her childhood years, Alyssa Miller has a passion for technology and security. She bought her first computer herself at age 12 and quickly learned techniques for hacking modem communications and software. Her serendipitous career journey began as a software developer, which enabled her to pivot into her cybersecurity career. Beginning as a penetration tester, she has spent the better part of two decades in cybersecurity. Those years have seen her grow as a security leader with experience across a variety of organizations and cybersecurity roles.

By the age of 30, Alyssa was already serving in leadership roles in cybersecurity. Running the security testing and vulnerability management program for a Fortune 200 financial services firm, she learned the rewards and challenges of hiring and developing cybersecurity professionals early on. During her years in consulting, she was called upon to build an application security consulting team from the ground up. Now as an executive leader for another large financial services firm, she yet again is putting the experience of building teams to work as she builds out a cybersecurity practice for a multibillion-dollar division.

Alyssa is a highly recognized and regarded professional in the cybersecurity industry. She regularly advocates for improving the security community, both in terms of professional disciplines and the workforce. Alyssa is often called upon to share her research with business and industry audiences through a wide variety of international public speaking engagements, interviews, and other media appearances. She uniquely blends technical expertise and executive presence to bridge the gap that can often form between security practitioners and business leaders.

Alyssa's goal is to change how we look at the security of our interconnected way of life and to focus attention on defending privacy and cultivating trust. She is a committed activist within the security industry. Her advocacy for the security community is also displayed in her work with various industry nonprofit organizations, board membership for multiple community conferences, and her membership in multiple executive leadership communities.

about the cover illustration

The figure on the cover of *Cybersecurity Career Guide* is "Jacout" or "Yakuts," taken from a collection by Jacques Grasset de Saint-Sauveur, published in 1797. Each illustration is finely drawn and colored by hand.

In those days, it was easy to identify where people lived and what their trade or station in life was just by their dress. Manning celebrates the inventiveness and initiative of the computer business with book covers based on the rich diversity of regional culture centuries ago, brought back to life by pictures from collections such as this one.

Part 1

Exploring cybersecurity careers

Whether you have just finished school and are entering the workforce or you are looking to make a change from your current path, launching a new career journey can be daunting. The cybersecurity career field has grown significantly in popularity both because of its importance to our world today and because it offers a hot job market. However, what really is cybersecurity? What career paths are available to someone seeking to launch a cybersecurity career? What skills do you need, and how do you go about choosing the right job?

These are all questions we will examine and answer in the first three chapters of this book. Chapter 1 defines *cybersecurity* and explores its history. We will talk about the role that cybersecurity plays in our world as well as some of the ideology often seen in the industry. In chapter 2, you'll get a better idea of the breadth of cybersecurity roles as we talk about some of the more common jobs in this field. We will discuss the characteristics that are most important for being a successful cybersecurity professional. Then chapter 3 describes key challenges and obstacles that make it difficult for some to enter this career path. We will map out the types of career progression you can commonly expect and dive deeper into the skill sets that you may have and may need to develop.

So get ready—you're about to discover the exciting, challenging, and often dynamic landscape of cybersecurity. By the time you finish part 1, you will have a far greater understanding of the road that lies ahead for you.

This thing we call cybersecurity

This chapter covers

- Defining the term *cybersecurity* and understanding its history
- Identifying the role, values, and ideology of cybersecurity
- Realizing the importance of diversity as we seek to improve cybersecurity

So, you want to help secure this new digital world we live in by starting a career in cybersecurity? If you have researched how to start a career in a security-related field, you've probably heard and read plenty of discussion about the cybersecurity skills gap. Maybe you've even seen studies suggesting that as many as four million cybersecurity jobs could be unfilled. However, if you're out looking for your first role, you're likely among those who've been on the job hunt for more than six months.

If you're nearing graduation or looking to make a career change, you've probably asked, "How do I get started in cybersecurity?" Unfortunately, if you've gone looking for that answer, you've likely discovered that no single generally accepted answer exists.

As a cybersecurity professional with over 15 years of experience, I've hired some terrifically talented people into their first cybersecurity roles. I've watched teams I've built blossom from humble beginnings into powerful and effective cybersecurity groups. Yet for all the success I've experienced in hiring and developing talent, I've also watched the security community struggle to define a clear career path from entry level to advanced roles. I've witnessed the worst in hiring processes, bad advice for beginners, and gatekeeping by long-established professionals.

The good news is you've purchased a copy of this book. In the pages to come, I'll help you understand the unique nature of what is commonly referred to as the *cybersecurity industry*. I'll take you on a journey that starts by defining the field you're looking to become a part of. I'll use interviews with various members of the cybersecurity community to demonstrate how seemingly unrelated skills and backgrounds can be an asset to a security career. I'll leverage surveys I've conducted of over 1,500 cybersecurity professionals and aspiring professionals to analyze paths you can follow to help speed your transition into a security role.

Over the course of this book, I'll analyze the value of education, training, certifications, and mentorships in landing a job. I'll share insights on how to interpret job postings for security positions and how to analyze and emphasize your unique experience to best position yourself to get hired into that first role. I'll give you a glimpse of the types of interviews that are typically used in the hiring process and share techniques for maximizing your performance. I'll even share my insights on how to ensure your continued success in your chosen career path after you've landed your first job.

The first step in the process of getting you that cybersecurity job is to understand what cybersecurity is, what the roles within cybersecurity are, and how they apply within different contexts of our daily lives.

1.1   What is cybersecurity

Cybersecurity is a term that has become ubiquitous in modern society. From news media, to politics, to the business world, cybersecurity is a topic that comes up daily in most people's lives. For all this discussion, however, it can be quite difficult to find a definitive answer to the seemingly simple question: what is cybersecurity?

No single generally accepted definition exists. Most will agree, however, that cybersecurity is an extension of what is often still referred to as *information security*. In 1961, researchers at the Massachusetts Institute of Technology (MIT) created the first password-protected system known as the *Compatible Time-Sharing System* (*CTSS*). For many, this is considered the birthplace of *information security*, which is the practice of protecting information and the electronic systems that process it from unauthorized access.

Fast-forward about a decade from those early days, and researchers were beginning to connect computer networks to the Advanced Research Projects Agency Network (ARPANET). This network was designed to allow other computer networks across wide geographic areas to communicate and share data quickly and reliably. ARPANET, as it turns out, would be the beginning of what we know today as the internet.

In 1988, however, three years before the internet was made available to the public, a researcher named Robert Morris wanted to highlight security risks in research computers that were connected to the internet. He designed a piece of software that spread itself across the computer systems connected to the internet. The software used security flaws in the UNIX operating system to install itself and then continue replicating. For all intents and purposes, Morris had created the first internet worm. Unfortunately for Morris, the worm spread out of control and made the infected systems unusable. This not only resulted in Morris being the first person convicted of a felony under the Computer Fraud and Abuse Act of 1986, but also led to the creation of the *Computer Emergency Response Team* (*CERT*) at Carnegie Mellon University under funding from the US federal government.

The creation of CERT can be looked at as the birth of what we now call cybersecurity. Therefore, a reasonable working definition of *cybersecurity* is the domain of research, technologies, and practices used to protect connected technology systems, data, and people from attack, unauthorized use, and/or damage.

1.2 *The role of cybersecurity*

The objectives of cybersecurity shift significantly depending on the context in which it is being applied. When cybersecurity is talked about in the media, it is often from the perspective of protecting business and commerce from cyber criminals and attackers. However, almost as common are discussions of how cybersecurity is applied across society at large. From securing our elections, to national/international security, to individual online privacy protection, cybersecurity is the common thread responsible for ensuring that all aspects of society function without disruption.

A solid understanding of the breadth of the cybersecurity world begins with understanding how cybersecurity fits into these various aspects of our lives. Cybersecurity has become so ingrained in everything we do that it can often be taken for granted or overlooked altogether. Taking a step back and examining in detail some of the diverse ways in which cybersecurity is relied upon will enable a stronger discussion when it comes to the disciplines and even job roles that are a part of this domain.

1.2.1 *Cybersecurity in the business world*

In business organizations, the goal of cybersecurity is typically to protect the company's financial interests. Organizations operate on a model of *assets*, the elements of the business that hold or create financial value, and *liabilities*, elements that decrease or carry a risk of decreasing financial value of the business's assets.

From the mid-twentieth century, *information technology* (*IT*) has been adopted by businesses to enable faster and more advanced capabilities. IT is the use of digital systems such as computers to manage and process information assets of a business. As IT systems have evolved, especially developments over the last decade, more and more business assets have become a part of the digital domain.

The term *digital transformation* has been adopted to describe this phenomenon of businesses digitizing their critical assets and becoming more reliant on IT systems. For example, health records that used to be stored in paper files and in images on physical film have increasingly moved to *electronic medical record* (*EMR*) systems. Storing all that information digitally in computer systems makes it easier to access, view, and share. In fact, an entire marketplace of IT products and services has formed around this digital transformation, to assist organizations making these conversions in just about any industry—from healthcare to education to transportation.

As businesses transform their assets to the digital realm, the risk of cybercriminals attacking those systems for those assets increases. These threats of attack can range from attempts to steal data, to attempts to make the systems unavailable for use. Information assets that at one time were at a low risk of being attacked now in the digital realm face the risk of attack from threats around the globe. The connectivity and immediacy of data access and interactions across the internet have enabled an explosive growth of assets in the digital domain, but have also enabled the emergence of new threats to those assets.

Cybersecurity technologies, practices, and resources are in turn relied upon to ensure that the risks posed by those threats are minimized. So cybersecurity's primary objective within business then becomes defending this ever-growing landscape of digital assets.

As discussed previously, businesses operate under risk management models to ensure their overall success. Leaders of companies large and small are always weighing the risks that the business could be negatively impacted by an event or shift in conditions, and then trying to minimize those risks. For example, an organization like Facebook may have to weigh the potential revenue from selling user data to a partner versus the potential liability of violating privacy laws. In addition, a business and its leaders must consider the potential cost if a threat successfully impacts an asset versus the cost of reducing that risk. These are complex decisions that drive financial decisions as well as other organizational strategies. So as digital assets become more a part of this landscape, it really is no surprise that cybersecurity would be subject to those same forms of risk analysis.

In this way, cybersecurity becomes a crucial input to the risk management process within an organization. Cybersecurity practitioners are often looked to for their expertise in assessing the level of risk to specific business assets from the various threats that could target those assets. This creates responsibilities for security teams that go beyond just technological capability. Security staff must be able to understand the threat landscape and effectively communicate the characteristics of those threats to other areas of the business that don't have the same level of technical knowledge. We need to be able to explain threat actors in terms of nation states, hacktivists, internal threats, and so forth, that are all a part of the threat landscape. Security staff must also be able to understand how assets fit within the overall business in order to more accurately describe the risks that threats pose to the business.

Since IT systems have become such an intrinsic part of the business model, their criticality to businesses has increased as well. A failure of a system that makes it unavailable for use can have enormous impact on a business. Think of some of the nation's biggest retailers and how much it would cost them if their cash register systems were unavailable even for a half hour. Healthcare facilities, financial institutions, logistics companies, and just about every industry imaginable has become reliant on IT systems to keep their businesses running.

Because of the criticality of these systems, which in our modern age are typically interconnected in some way, cybersecurity also plays a role in ensuring the stability and availability of those systems. Attackers seeking to do damage to an organization might attempt a *denial-of-service (DoS)* attack, trying to make the business's systems inaccessible for a period of time. Cybersecurity professionals are tasked with preventing the success of these types of attacks as just one of their items on a long list of responsibilities.

Typically, this type of defensive approach is done in conjunction with a team that is primarily responsible for the day-to-day ongoing functioning of the systems. In IT, these teams are typically referred to as *operations teams*. As it applies to cybersecurity, teams that focus on the day-to-day functioning of security defenses are referred to as *security operations teams*.

> ### Examples of day-to-day responsibilities in cybersecurity
> The following are some of the typical responsibilities of cybersecurity professionals:
>
> - Monitoring for attacks across various systems
> - Responding to successful attacks that breach a system or systems
> - Assessing systems and people for security weaknesses (known as *vulnerabilities*)
> - Tracking, validating, and reporting on the fixing of those vulnerabilities
> - Working with developers on practices for developing secure software
> - Designing and deploying security measures (also known as *controls*)
> - Working with executive leaders to secure budgeting for security
> - Providing evidence of security controls for auditors
> - Maintaining various security systems (user accounts, firewalls, and so forth)

As business models become more heavily dependent on digital assets and IT systems, yet another trend has emerged. The level of government regulation and industry compliance requirements surrounding the use of IT systems has grown at a breakneck pace. Many of these regulations and compliance standards include detailed requirements for the way organizations secure their systems, respond to breaches or data exposures, and go about protecting consumer privacy.

Once again it is no surprise, then, that the cybersecurity employees within an organization play an important role in the way the company achieves, maintains, and demonstrates compliance with these various regulations and standards. To begin with, the security personnel are often called upon to digest and even interpret what the requirements actually mean. This may be done in collaboration with other areas of the business such as the legal team, risk management team, or audit team, but the expertise that security brings to those discussions is crucial.

Following this interpretation, security expertise is needed in designing and implementing the various controls that will ultimately ensure the organization's compliance with these requirements. These controls can take the form of processes, practices, policies, and technologies that are all intended to help the organization protect its data and systems sufficiently according to the requirements.

Looking at the role of cybersecurity within a business setting, it becomes clear that security personnel have become involved in just about every aspect of the business. Whereas traditional information security teams were often able to focus exclusively on technical IT access controls and countermeasures, the modern digital world has forced security to be a part of every business conversation.

1.2.2   *Cybersecurity defending society*

Moving from the business world to the broader perspective of society changes the focus of security professionals. As intertwined as cybersecurity has become in the day-to-day motion of conducting business, it is equally or even more so a regular part of our everyday lives. The functioning of our government, our national security, law enforcement and crime prevention capabilities, and even personal interactions have all come to depend on the digital realm within our twenty-first-century society.

All levels of government have become incredibly reliant on computer and mobile applications, digital data, and other technological capabilities that are part of the digital world. If there is any doubt about just how important IT systems have become in the daily functioning of our government, we need only look at *ransomware attacks*, in which malicious software is installed on a computer to make the data unavailable until a ransom is paid to the attackers.

One of the more notable attacks against a local government happened in Baltimore, Maryland in May 2019. Portions of the city's government were shut down, some for more than a month, as email, payment, and other systems were suddenly unavailable. The lost revenue plus recovery efforts cost the city over $18 million. Many other local, state and national governments around the globe have experienced similar attacks.

Of course, daily functions are not the only way that the government relies on IT systems. The use of electronic systems to handle voting is also growing rapidly. With the public demanding faster and more accurate access to results, governments across the United States and around the world are turning to digital voting terminals. However, the threats to these voting terminals have also been well-documented. Security issues and potential hacking attempts have been identified in past elections, most notably the 2016 and 2020 US presidential elections. Ultimately, the US

Cybersecurity and Infrastructure Security Agency (CISA) and independent security firms all concluded that, thanks to the efforts of cybersecurity professionals, no attempts to hack those systems were successful.

Security professionals and researchers are regularly sought after by government agencies for help in defending against attacks. The stakes couldn't be higher. Little within the government space can be considered low risk if it is impacted by a cyberattack. Even when parks, museums, or other government-managed services are affected by an attack, the negative public reaction can be swift and powerful. No political candidates want their name attached to a cyberattack occurring on their watch. As a result, momentum is growing for concerted efforts—which many security professionals would say are overdue—to shore up security within government agencies.

But the problem extends beyond civilian government matters. Militaries around the globe have also become increasingly dependent on technology systems in their efforts to defend their nations and those of their allies. Everything from military vehicles to communications to monitoring systems leverage increasing levels of connected technology. Beyond any other application, cybersecurity within the military is at the peak of life-and-death significance. As new technologies are introduced, governments and their contractors turn to security researchers and practitioners to help ensure that those systems are sufficiently protected against attacks, from design through their use in the field.

A natural extension of military use is the enforcement of laws at a domestic level. From active patrols and dispatch to investigations and criminal justice, computers and other connected electronic devices play a key role. Attacks against these systems could have detrimental effects on the departments they serve and make enforcing laws and prosecuting violations of those laws impossible. Additionally, given the ever-growing interconnectedness of our society, many crimes are committed using electronic means. Having skilled security professionals to not only defend the department or agency's systems but also assist in investigating crimes is vastly important.

Finally, the daily lives of individual citizens around the globe are completely intertwined with connected technology. From social media, to electronic communications, to mobile apps and even so-called smart devices, human beings on this planet have largely become inseparable from technology. This creates an ever-growing pool of targets for cybercriminals to attempt to exploit. Many who use these technologies are unfamiliar with practices for using them securely and not exposing themselves to attack. As a result, security researchers and professionals are looked to for their expertise. Whether it's through increasing awareness or developing and implementing countermeasures or even identifying security vulnerabilities in consumer electronics and software, cybersecurity is looked to as the answer for protecting every person on the planet who is connected through technology in some way.

1.3   The cybersecurity culture

For decades, a community of people committed to goals of deconstructing, investigating, and defending technology has been growing and evolving. This community has developed a culture and many subcultures that have shaped much of cybersecurity's structure today. From hackers and researchers to security practitioners and corporate security leaders, a unique and sometimes difficult-to-navigate set of norms and values have come to be associated with the security community.

It would be impossible to list every core value or ideology that has been adopted by the security community. They not only are far too numerous, and in some cases ethereal, but also are not universally adopted by all who would identify as members of the security community. However, several values are widely held that should be examined to provide better context for anyone trying to become a member of the community.

1.3.1   Privacy and liberty

Key tenants in the ideology of those within the security community are personal liberty and privacy. In the early days of hacker culture, individuals around the world gathered on dial-up server communities (known as *bulletin board systems*, or *BBSs*) to share information and discuss new discoveries. To gain access to these systems, participants often had to demonstrate proof of a "hack" they had conducted.

That often meant showing data they had stolen from a business whose systems they broke into or demonstrating that they were able to manipulate other technology to cause it to function in a way that wasn't intended. Since these activities were often viewed as illegal, the ability to protect their personal identity and remain free from watchful eyes of governments and officials by maintaining anonymity was highly valued.

A significant portion of the members in these communities were treated as outcasts in their daily lives. What they found in the anonymity of these early communities is described by many as a feeling of being among people like themselves. Stripped away were labels of gender, ethnicity, social class, or other ancillary characteristics that led to rejection from mainstream society. Instead, each was valued almost exclusively based on the knowledge and skills they brought to the table. They could have meaningful discussions about topics they wanted to discuss with others who had similar interests without stereotypes or prejudices getting in the way.

As the internet began rising to prominence, the early design and capability limitations of internet technology enabled continued anonymity plus greater convenience in connecting to vast communities of like-minded individuals. However, the secretive, often clandestine nature of these early hacker groups in many cases began to erode. They became more visible to the general public, and interest in their activities grew.

At the same time, as discussed earlier, within corporations and government agencies, the ideas and practices of information security were also growing. Industry, law enforcement, and government groups that focused on information security practices began to cultivate their own communities of security professionals.

Over time, these two very different groups of individuals have developed a tenuous, if not strained, relationship. Through meetups, organizations, and even formal security conferences, the two groups have found ways to share information, ostensibly with the common goal of making technology better and safer for all. It's the ideological view of what makes technology "better" that often still differs between these groups.

This leads to a continuing distrust and sometimes outright animosity between the two factions. As a result, protecting privacy and liberty has been reinforced as a value particularly among the more idealistic hacker/researcher portion of the community. Still today, many in the community use *handles*, nicknames meant to protect the actual identity of the person and operate under general anonymity.

1.3.2 Open information sharing

One of the key elements that brought early hackers together was the ability to *share information freely* with one another. These hackers weren't the cybercriminals we hear about today; they were simply people who sought to better understand technology so they could learn and create even more innovative technology. However, this information sharing was sometimes accompanied by a level of hubris. Bragging about a recent hack meant greater credibility in the community. Regardless, the value of sharing information and building on others' discoveries was and is important to the community.

This type of information sharing isn't exclusive to the hacker culture. Academic researchers also have long valued the concept of open information sharing, and indeed that continues into cybersecurity research.

This culture of sharing information to help improve technology for the good of everyone is seen, in particular, in the number of independent security conferences hosted annually. Thousands of conferences are held around the globe in the interest of sharing information about security vulnerabilities, defenses, and other topics. A week-long series of security-focused conferences, colloquially referred to as *Hacker Summer Camp*, takes place in Las Vegas each August and attracts an estimated 30,000 to 40,000 people from around the globe.

The importance of this ideology within the security community is manifested in the way its members have reacted to the commercialization of the internet. In its infancy, the internet was a new frontier that would enable the free sharing of knowledge on a level never before possible. For a time, that ideology seemed to be holding true. However, it didn't take long for businesses to realize they could enable new revenue streams and reach customers in a way never possible by leveraging the internet.

To protect competitive advantages and establish markets, businesses maintained their corporate secrets even as they exploited more of the capabilities of the internet. New technologies that enabled more secrecy and controversial defense of certain intellectual property rights conflicted with this open information-sharing ideology. The security community in turn has continually fought to tear down those barriers and gain more transparency from businesses in terms of their business practices on the internet.

1.3.3   Do no harm

Early hackers quickly understood that their capabilities could be used to improve the quality of technology for all. As they began discovering security flaws in systems, they sought ways to share this information with the owners of those systems. Unfortunately, those system owners and ultimately law enforcement viewed the activities of these hackers as criminal rather than helpful.

Slowly, however, businesses and even law enforcement agencies began to realize that understanding the perspectives and skills of friendly hackers could help them defend against the actions of truly malicious attackers. From this, the term *ethical hacker* emerged, describing someone who used hacking techniques to help discover security flaws for the purpose of reporting those flaws so they could be fixed. While this term has fallen out of favor, the concept is alive and well.

To aid in establishing legitimacy, the ethics of friendly hacking needed to be carefully established and adhered to. This allowed good hackers to define rules, practices, and standards that differentiated them from malicious attackers. Imperative to this ethical code was the ethos of *do no harm*. This formalized rules of engagement for testing systems, ensuring that while vulnerabilities would be discovered, they would not be exploited in a way that caused damage to a system or a person.

In today's cybersecurity world, this code lives on and is applied to many of the activities that security researchers, hackers, and practitioners engage in every day. Debates rage when ideas of offensive security and cyber warfare seem to cross the line and inject truly harmful behaviors in the name of protecting security.

1.4   The cybersecurity "industry"

Whether it's in the business world, in the media, or in political discourse, the term *cybersecurity industry* is often used to describe the full collection of people, technologies, and practices that are part of defending the digital world. Security, from the early days of information security professionals, has been viewed as a separate discipline.

We have this concept of *cybersecurity careers*, which is likely why you're reading this book in the first place. Governments, corporations, and other entities have built *cybersecurity teams*. Software and hardware companies have released *cybersecurity products* to try to defend against every imaginable type of attack. But uncertainty is growing over whether cybersecurity should be looked at as a separate industry at all.

1.4.1   Is cybersecurity an industry?

Cybersecurity has been well established as a commercial market. Various studies indicate that $170 billion to $250 billion was spent globally on cybersecurity solutions in 2019. Additionally, colleges and universities have created degree programs that focus on cybersecurity. Training organizations offer cybersecurity bootcamps and classes. For years, as information security teams stood as lone silos within corporate organizational charts, viewing security as an industry was convenient and made sense.

However, cybersecurity has evolved into more than simply protecting IT systems from unauthorized access and damage. The implementation of its practices is no longer solely a matter of technical countermeasures. The focus on security has permeated into every area of business, international affairs, and societal dialogue. Calling cybersecurity an *industry* connotes a standalone silo—something that exists and merely interacts with other facets of our world. That connotation fails to recognize that security is a fundamental concept in every part of the digital world.

> **DEFINITION**   Cybersecurity is more than an industry. It is important to understand that cybersecurity is intertwined in every facet of our digital world. Therefore, referring to it as an *industry* perpetuates an antiquated view of security as a siloed and separate function within organizations and society.

1.4.2   *The effects of digital transformation*

As discussed in section 1.2, digital transformation has resulted in the conversion of many elements of daily life to an electronic and digital realm. Technology is no longer a part of our lives; in the words of a colleague and good friend in the security community, Keren Elazari, it *is* our way of life. We're no longer just defending systems, technology, and data, but instead are defending core aspects of our modern world.

As digital transformation continues, and more and more once-tangible elements of the world around us become digitized, cybersecurity becomes ingrained in that aspect of the world. Threats are growing exponentially in number and in complexity. No single group, no single discipline, no single domain of expertise can reasonably be called on to defend against all of it. The resultant attack vectors are too diverse and expansive.

1.4.3   *The human element*

The growth of our digital world through the transformation of everything we know to data and systems has highlighted another key concept: the need to protect the *human element*. An often-cited idea within the security community is that the human element is most often the weakest link. No matter how strong our defenses, no matter how good the technology, a single mistake made by a human being can still enable a malicious actor to complete an attack.

As this concept has grown, we've seen the introduction of social engineering experts into the cybersecurity disciplines. Organizations pay these practitioners to assess the readiness of their personnel to defend against attempted attacks. Experts focused on human behaviors and awareness training have become crucial elements in this type of security strategy. These experts focus on reshaping the way humans react to attempted social engineering attacks such as phishing, phone fraud, or even in-person manipulations. In 2020, the RSA Conference (one of the largest and longest-running cybersecurity conferences in the world) featured "Human Element" as the theme for its annual week-long event in San Francisco.

The inclusion of defending the human element against attack further broadens the idea of what cybersecurity is. It forces practitioners to think beyond just technology and really consider inherent behavioral patterns, manipulation techniques, and disinformation countermeasures.

1.4.4   *The internet of everything*

Digital transformation has changed much about the way we view our world in ways we're only just starting to understand. One example is the *Internet of Things* (*IoT*), more commonly now referred to as *smart devices*. Both terms describe products and technology that typically operated in a standalone fashion and now are augmented with connectivity to create a new form of functionality.

Refrigerators can detect when items are running out and order more from an online grocery store. Cars are connected to the web for wide-ranging purposes, from navigation assistance to summoning assistance when needed. In February 2020, a Kickstarter campaign was even announced for a candle that could be remotely lit by using an app on a smartphone. Everything seems to be getting connected to the internet.

However, this explosion in connected devices has predictably also caused explosive growth in threats and attack vectors. Security considerations are now a part of technologies that were never thought to have a digital threat landscape previously. Once again, security is permeating into every facet of our lives.

So, can we really refer to security as an industry anymore? Should we instead look at security as simply a facet of every part of our world, the way that safety has been for generations? Sure, some people specialize in designing safe environments. Best practices exist, standards have been created, and concepts are followed. But in the end, from workplaces to roads to homes and everything in between, safety is just an inherent aspect of all of it. Perhaps as we consider career paths and specializations, thinking about cybersecurity in the same way could be helpful.

1.4.5   *So, is cybersecurity an industry?*

As you can see, the cybersecurity field itself is broad and extends across not just the ever-expanding world of technology but also to securing people. Our way of life, in almost every facet, has become integral with this digital world.

So, to say that cybersecurity is just an industry ultimately is far too limiting. It is a crucial element in the way we approach day-to-day life and not something that is easily separated anymore. Whereas information security in past decades could be looked at as just a discipline under the larger information technology umbrella, cybersecurity today is more conceptual and less of a specific skill or set of practices.

1.5   *The value of human diversity in cybersecurity*

In section 1.4.3, we discussed the human element and how the efforts we make within cybersecurity to defend our way of life with technological means can be undone by human mistakes. Therefore, the problem-solving that sits at the core of what we do in terms of cybersecurity must include those humans whom we ultimately seek to defend. But with a world that has so many cultures, so many differing sets of ideals, and so many levels of education and abilities, how can we hope to find answers to defending all of them?

As it turns out, one big step that must be taken toward this end is ensuring the diversity of those who are called upon to build those defenses. To protect our digital way of life, we need to improve our problem-solving through diversity of thoughts, perspectives, and ideas. Further, we need to understand the populations that we are trying to protect.

Neither of these can be accomplished if our teams of defenders all have similar backgrounds, similar educations, similar cultures, and similar career progressions (among others). To make this idea of cybersecurity work, we have to be welcoming and actively seek to include defenders from walks of life that are just as varied as the societies we live in. This means there is a place for all in cybersecurity roles. But furthermore, we truly need to have as much representation in these roles as possible.

1.5.1   *The cybersecurity diversity gap*

In its 2020 "Diversity and Inclusion Report" (http://mng.bz/PW92), cybersecurity company Synack surveyed hundreds of professionals about their experiences working in cybersecurity roles. The survey asked whether the respondents felt they were given the same opportunities to progress in their careers as those of other genders or ethnicities. Of the female participants, 34 percent answered that they did not. More alarmingly, 53 percent of those from minority ethnic backgrounds answered no to the same question. These results are indicative of the diversity problem that has plagued the technology industry, and particularly cybersecurity, for years.

In 2017, the International Information System Security Certification Consortium, or (ISC)², and Frost & Sullivan and others, released a "Global Information Security Workforce Study" (http://mng.bz/J1Op). The survey found that only 14% of respondents in North America were female. Across all other regions of the globe, that number was even smaller. A lot of press and analysis have focused on the issue of female representation in cybersecurity, yet the trouble persists. The same study found that only 23% of United States–based respondents identified as an ethnic minority, which also falls below overall population percentages for the nation. For instance, while 13.4% of the nation's population identifies as Black or African American, only 9% identified as such in this study.

Conflicting reasons are often shared to explain this gap. I will not debate the merits of those theories in this book. However, it is important to understand that these challenges in diversity do exist—in particular, gender diversity—and this has an impact on how successful we can be in our pursuits. It is equally important to understand that this problem has been recognized and that our community as a whole is working to change it.

1.5.2   *Why it matters*

Diversity is often touted as a political correctness or "woke" effort. But the technology world and cybersecurity are slowly coming to realize a tangible value in diversity that goes beyond just simple ideals of morality and fairness. As I stated at the beginning of this section, we need cyber defenders to understand the mindset and perspectives of those we seek to defend, especially when the human element is the cause of so many challenges in that regard. This ability to understand the humans to which our efforts are directed allows us to better identify the right solutions that will work in defending them.

By way of example, in 2014 the US Government Accountability Office (GAO) released a report (https://www.gao.gov/assets/gao-14-357.pdf) that detailed elevated instances of false alarms in Transportation Safety Administration (TSA) body scanners at airports. Among the particular associations with these elevated numbers of false alarms were headgear, turbans, and wigs. Yet in 2017, ProPublica reported (http://mng.bz/wnd7) that the scanners continued to have high levels of false alarms, especially associated with hairstyles common among African American and Black women. This was according to data that ProPublica collected independently.

In these scenarios, we have to wonder how such issues do not get identified sooner. Were women of color not involved in the development and testing of these devices? Could a more diverse project team have proactively recognized the potential for these issues and ensured that the designs of the system took this into account? Ultimately, this is why diversity is so important. Brainstorming and problem-solving benefits when those involved have wide-ranging perspectives and experiences to draw from. So when it comes to cybersecurity, where problem-solving is the core of our calling, we too must seek to have significant diversity within our community.

1.5.3   *How it applies to your career path*

This is all great, but it sounds like a problem for the community and industry as a whole. While this is true, as someone looking to launch a career in cybersecurity, it's important to understand that these challenges exist. In chapters 8 and 9, I discuss the various challenges that can derail your career growth. For now, this particular concept needs to be recognized early as you just start to understand the makeup of the cybersecurity community, how we got to where we are today, and where we are headed.

As you begin down this career journey that has led you to this book, you might struggle to see how you fit into the community if you do not see yourself represented in the faces of those already there. That is the point where you will need to draw upon the information in this section and understand that you are not only welcome but also needed. Armed with this information, in the next chapter we will really dig into all the places you could go within cybersecurity.

Summary

- Cybersecurity is the domain of research, technologies, and practices aimed at protecting connected technology systems, data, and people from attack, unauthorized use, and/or damage.
- Cybersecurity's role changes with context, but in our digital world of interconnectedness, it's about protecting our way of life.
- Cybersecurity can benefit greatly from diversity of experiences and cultures, and the community continues to work to improve the current lack of diversity.

The cybersecurity
career landscape

This chapter covers

- Various disciplines included in cybersecurity
- Profiles of cybersecurity professionals across these disciplines
- Characteristics of good cybersecurity professionals

In chapter 1, you began to explore just what cybersecurity is and why it even exists. Now it is time to dig a little deeper and explore the various career paths and disciplines that make up the modern day cybersecurity community.

You also saw in the first chapter how diversity is so important to our efforts to defend the digital way of life that we have come to rely on in society. As we now begin to explore the disciplines of cybersecurity, you will see just how diverse they can be as well.

2.1 The many disciplines of cybersecurity

Anyone considering a career in a cybersecurity-related field should take the time to understand the high-level disciplines that exist within this community. Each of these disciplines, depicted in figure 2.1, categorizes numerous job roles that are constantly evolving and changing. Understanding which disciplines best fit your abilities, experience, and interests will help you better focus your efforts as you seek to build a career. Given the dynamic nature of security specializations, an exhaustive list cannot be put together and kept current.

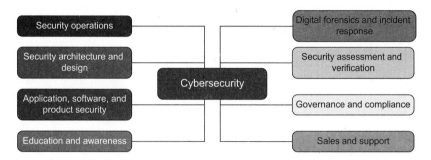

Figure 2.1 A map of the high-level disciplines in cybersecurity

Additionally, no universally accepted mapping exists of these roles and categories. Professionals across the cybersecurity community have made countless efforts to diagram and document the various roles. Since the technology and our approaches continue to change, so too do the roles and the way we choose to categorize them. However, the high-level disciplines that we discuss here will at least help capture the breadth of those specializations for ease of understanding. Additionally, we'll discuss the many leadership roles that sit across the top of these various disciplines.

2.1.1 Security operations

When we talk about security operations, we're talking about the people on the cybersecurity front lines. These folks are responsible for day-to-day systems management, monitoring, triage, and initial response to incoming attacks. The skill sets of these people tend to be wide ranging as they're ultimately responsible for maintaining security posture across all technologies within an organization.

The responsibilities within security operations therefore are wide reaching. At the core of security operations is typically the *security operations center* (*SOC*). Within many organizations, this group (and often location) is dedicated to constant monitoring of systems for signs of security-related events. When a potential event is detected, these resources must triage it to determine whether it represents an attack and then initiate the proper response when necessary.

However, security operations can extend beyond just the SOC. Often, administration of basic security controls within the organization fall under security operations. These roles might be filled by a security help desk or by individuals within a more federated structure. Ultimately, they're responsible for day-to-day maintenance activities like user account administration and desktop security software management. Additionally, within larger organizations, these types of administration and systems management roles may be handled by groups of specialists. For instance, those responsible for maintaining a particular defense technology may interact with the SOC and the help desk but bear the responsibility of actually maintaining and configuring those defenses.

The security operations role also interfaces with many of the other roles that we will discuss. For instance, when an alerted event turns out to be an attack, they may need to engage the incident response team to provide a more sophisticated response to the problem (more on that soon). They also need to be aware of the current threat landscape, because receiving regular information from threat intelligence resources is crucial.

For many who are looking to begin a career in cybersecurity, security operations is where they start. Security operations roles often work with automated systems and repeatable tasks that lend themselves well to on-the-job learning and training. Additionally, individuals in these roles can easily leverage previous IT experience in their daily job functions. Finally, because of the nature of the role and the wide-ranging responsibilities over various forms of technology, security operations is a terrific way to gain exposure to a lot of the technologies and concepts that cybersecurity teams are charged with defending.

2.1.2   *Digital forensics and incident response*

One of the key aspects of cybersecurity is, of course, responding to incoming attacks and investigating the origins and impacts of an attack that has occurred. The skill sets and roles focused on these types of activities are collectively known as *digital forensics and incident response* (*DFIR*). Whereas the SOC is responsible for evaluating potential incoming attacks and taking initial steps to defend against them, they will typically escalate to incident response personnel if a more coordinated and specialized level of response is needed or if the breadth of the attack involves extensive portions of the environment.

Incident response (*IR*) personnel are ultimately called upon to provide a coordinated and highly methodical response to security incidents. The criteria used to define a security incident can vary significantly from one organization to the next. Factors like risk tolerance, business criticality, and SOC capabilities often come into play in determining those characteristics. The process followed when an incident occurs can also be quite different depending on the organization. The organization's size, security capabilities, regulatory landscape, business model, and more can all be factors in the way an organization defines its process for responding to security incidents.

The criteria for what constitutes an incident along with the process for responding to an incident is often documented in an *incident response plan*. Having an IR plan should be best practice for all organizations. The plan is meant to be consumed by the entire organization. It provides the steps for reporting potential incidents, the IR team's response process, as well as other areas of the organization that should be involved and the processes that they should follow.

The IR team is typically responsible for managing and updating this plan. However, given that it touches every other part of the organization in some way, the IR team often works with groups including legal, marketing, public relations, the SOC, other areas of the security team, and even business and product development teams.

Since the IR team must be able to respond to incidents that could occur in any of the organization's technologies, many skills are useful for a role in incident response. Experience with networking, software development, wireless communications, and even human communications are crucial in the IR team. Organizations with mature IR functions understand the value of having diverse individuals on the IR team. In some organizations, the IR team might also be responsible for digital forensics, while in others, these two roles are separate.

Digital forensics refers to investigating the facts of a security incident (or even potentially nonsecurity-related events), often after the incident has been resolved. Digital forensics involves reviewing evidence found on systems to determine how an attack occurred and which data and systems it impacted. Digital forensics personnel are also responsible for identifying, documenting, preserving, and providing expert opinion on evidence that is collected from the various systems involved in an attack. Within highly regulated businesses, this role can be particularly critical.

As you may have guessed, all this means that a wide-ranging set of skills is needed for digital forensics personnel. Understanding low-level system behaviors can be crucial, but so can understanding the chain of custody and other evidence-handling concepts. Digital forensics personnel need to be particularly detail oriented and able to write effective documentation on a regular basis.

While DFIR roles don't tend to be entry level, experienced professionals from outside the security community can often still be well qualified for a role in DFIR. Former law enforcement professionals often find a niche in DFIR, given their experience with investigations and evidence handling. Network and systems administrators as well as infrastructure support personnel often understand the low-level behaviors and available event data within wide ranges of technologies. This understanding is also useful in a DFIR role. Additionally, communications and writing skills are absolutely crucial in this space. Sometimes those who do the technical tasks and those who document the results can be two different roles, although that is not particularly common.

2.1.3   Security architecture and design

Security architecture and design is a more proactive discipline within cybersecurity, but it is no less crucial than any other role. People in this field are responsible for specifying, designing, and typically implementing security controls and technologies. These roles are often held by individuals who are familiar with a wide range of security technologies, from firewalls and end-point protection to *security incident and event management* (*SIEM*) solutions.

These roles can interface with a lot of areas across the organization. The technologies they specify need to be appropriate and usable at a wide scale. Within large companies or government agencies, that can be particularly challenging. Personnel in these roles need to be able to interact with nonsecurity and often nontechnical roles to ensure that their designs and security architecture meet the needs of the business.

Security architecture and design resources also need to have particular familiarity with emerging threats and be able to at least informally develop threat models. They need to be able to understand the threats the organization faces, prioritize them according to the business's overall risk landscape, and then develop security solutions that sufficiently address or mitigate the risk posed by those threats.

As with so many security-related roles, communications skills are also crucial. The security architect is responsible for not just defining requirements and designing a security solution, but also communicating that solution to various levels of the organization in order to win funding and support. This can be a particularly challenging aspect of the role, as it often requires bridging a gap between technical expertise and business acumen. The architect needs to have the technical skills to design effective controls, but also understand business motivations and practices to effectively motivate senior levels of management to support the initiatives the architect is proposing.

Security architect roles require significant and diverse experiences. As a result, they can also be some of the most highly compensated roles within the security organization. Being able to draw upon a wide range of past experience with varied technologies, security approaches, and frameworks, while also staying current on emerging threats and trends, is critical to success as a security architect. These roles are hard for organizations to fill but are effective in improving the security posture of the company or agency.

2.1.4   *Security assessment and verification*

One of the more popular and well recognized areas of cybersecurity is the security assessment and verification discipline. When people think of cybersecurity, the first image in their minds is often that of a penetration tester (previously referred to as an *ethical hacker*). While this role does fall within assessment and verification, the discipline comprises many roles beyond just penetration testing.

Security assessment and verification includes the domain of activities that focus on confirming the security posture of an organization by attempting to identify and eliminate security weaknesses. This includes activities that not only seek to find vulnerabilities but also to provide tracking of known weaknesses to ensure they're properly prioritized and addressed. Within IT, the scope of these activities can include networks, infrastructure components, and software. However, the scope also extends beyond just IT systems. It can also include physical security, personnel, and even a review of publicly available information about the organization. The latter is also known as *open source intelligence* (*OSINT*).

Several activities can be a part of the security assessment and verification practices within an organization. Table 2.1 describes these activities and their characteristics.

Table 2.1 Common security assessment and verification activities

Activity	Description	Goals
Vulnerability scanning	Periodic automated scanning of an organization's environment for security weaknesses.	Provide regular security verification across a wide breadth of the IT infrastructure.
Penetration testing	Assessing the defined scope of IT systems by using attacker-style techniques usually consisting of both automated tools and hands-on technical hacking.	Gain a comprehensive view of the security vulnerabilities within the target systems with deeper capability and intelligence than a vulnerability scan.
Red teaming	Attacking a defined scope of IT systems in an attempt to achieve a specific objective by using attacker-style techniques while evading detection.	Emulate what a real attack may look like for understanding the true risk presented by vulnerabilities in the environment.
Purple teaming	Similar to red teaming, but adding collaboration with the teams responsible for detecting and defending against an attack (the blue team).	Emulate what a real attack may look like to determine and refine the ability of security controls to repel or recover from an attack.
Social engineering	Using deception and human psychology to attempt to convince personnel to expose sensitive data or provide access to sensitive resources. Usually done via simulated phishing and/or vishing attacks.	Determine how personnel within an organization respond to attempted manipulation and to build greater awareness of proper security practices.
Physical security	Attempting to bypass physical controls without being detected to achieve access to sensitive areas or items. Typically includes social engineering components (often in-person interactions) as well.	Emulate attempts to gain unauthorized access to facilities in order to validate that physical controls and personnel responses can sufficiently repel such attempts.
OSINT audits	Investigating public sources of information, looking for sensitive information about the organization or its people.	Evaluate whether sensitive data is being exposed in the public domain by business practices or personnel.

The activities involved in assessment and verification represent a wide range of skills. When we look at penetration testing, red teaming, and purple teaming, clearly a great deal of technical knowledge is needed. These roles require the ability to identify indications that a vulnerability may exist and an understanding of exploitation techniques to verify and potentially exploit discovered vulnerabilities. Since the target systems may be network devices, operating systems, systems software, or application software, experience in those areas is particularly helpful. For instance, many former software developers tend to specialize in performing application-focused penetration tests, as their experience in programming allows them to more easily recognize signs that secure coding practices have not been followed.

Social engineering, by contrast, doesn't necessarily require deep technical skills. Instead it requires knowledge of human interaction principles, psychology, and other human behavior domains that can assist in successfully influencing people to perform the desired tasks. Previous education or experience in psychology or even sales can often provide useful skills that are applicable to social engineering. Many of the qualifying skills of a social engineer aren't so easily quantified. It takes a special type of person, someone who is perceptive, can think quickly on their feet, and in general has strong interpersonal skills. However, the contribution of technical acumen can't be completely discounted. Social engineers need to be able to collect information about their targets, and certain technical capabilities can be helpful in that respect.

This linkage of technical and nontechnical skills goes even a step further when we start talking about physical security assessments. All of the preceding social engineering skills are certainly required since in these assessments it is common to have to interact with and influence security guards, employees, and other individuals. However, unique technical capabilities are also needed. Lock-picking, in-depth knowledge of electronics, and even radio communications experience can prove valuable to the physical assessment engineer.

However, as described earlier, this discipline goes beyond simply identifying security weaknesses. A crucial component for any organization is *vulnerability management*: the processes and systems used to catalog, prioritize, and ensure remediation of security vulnerabilities within an organization. Some may argue from an idealistic sense that this particular role fits within security operations. However, in practical application, more often than not, vulnerability management is connected most closely with the assessment and verification teams. In some cases, vulnerability management may be part of the responsibility of a corporate risk management team. This team looks at all forms of risk to the organization and helps leaders understand which risks are most important to be addressed. In this sense, vulnerability management is a good fit as a specialization within that broader team. Ultimately, it is the assessment and verification team that is tasked with finding weaknesses so they can be fixed before the "bad guys" find and exploit them.

To further demonstrate just how widely varied the careers in this particular category can be, I spoke to Kwadwo Burgee, a senior vulnerability analyst for the Cybersecurity and Infrastructure Security Agency (CISA), a US government agency that falls under the Department of Homeland Security (DHS). Burgee is responsible for acting as a broker between researchers who find security vulnerabilities in widely used software and products and the organizations that are responsible for maintaining those products. In particular, he is responsible for helping manage the Common Vulnerabilities and Exposures (CVE) database (owned by the nonprofit organization MITRE). Although he is not necessarily assessing networks and software, he and his team need to understand these concepts as they help foster the reporting of identified security flaws. Among other tasks, his team is responsible for ensuring that vendors to whom a flaw is reported provide an adequate response to those reports.

2.1.5 Application, software, and product security

Application, software, and product security focuses on elements that are developed by the organization for sale to customers, to support services provided to their customers, or to support internal processes. Application security commonly describes how organizations ensure the security of software they've developed. Software security is used a little more ambiguously but mostly as a way to describe security around any form of software used within the organization, whether internally developed or purchased externally. *Product security* refers more generally to securing any products, whether they're software or hardware based, that a company sells to its customers. The common thread across all three is that a life cycle is applied to the introduction of these elements, and the goal is to inject security practices into those life cycles.

Application security is well understood as applying security practices within the *software development life cycle (SDLC)*. Studies over multiple decades have repeatedly shown that finding and fixing security flaws early in application development is cheaper and more effective than if the flaws aren't discovered until the software has hit production.

Since it was introduced in 2008, DevOps has challenged security to remain an effective part of the software delivery pipeline. In a *DevOps culture*, software developers and operations support personnel work collaboratively within a shared responsibility model. As security practitioners have sought to keep security practices integrated into the early stages of software development, the idea of *DevSecOps* has emerged. The security practices within DevSecOps are typically addressed within the application security discipline as well.

Software security expands on application security by acknowledging that software that was not developed by the organization may also be used by the organization. Commonality between the life cycle of third-party software acquisition and deployment, and internal software development, are addressed within software security.

Product security is typically a security practice involving products that are sold by the organization. Sometimes this is a way of differentiating that the products are not necessarily software or applications. They could be devices or other tangible products that the company needs to make sure are secure. Again, since a life cycle is applied to developing and updating these products, security needs to be integrated into that life cycle.

Technical skills are important within the application, software, and product security discipline. Understanding how software and products are developed, down to a code or component level, is extremely beneficial for these roles. In many cases, it could be said it's even mandatory. It would be difficult for a cybersecurity practitioner to work with developers on secure coding practices if the practitioner had never written any code. Additionally, the empathy from having worked in a similar environment is helpful when attempting to influence the behaviors of software or product development.

2.1.6 *Governance and compliance*

Earlier in this chapter, the idea of *governance and compliance* was first introduced. Organizations in both the public and private sector are increasingly regulated by industry and legal requirements. Additionally, most organizations have internal policies and standards that describe required security controls, processes, and technologies. It is the job of governance and compliance personnel to ensure that all areas of the organization are compliant with external regulations and internal policies.

Typically, governance and compliance is a role that is much bigger than just cybersecurity. Corporate policies as well as industry and government regulations typically extend to more areas of business concern. However, as more privacy- and security-related regulations are being enacted, the need for skilled individuals with cybersecurity knowledge is growing.

Governance and compliance personnel work with more of the administrative side of the organization. Interaction with legal and audit functions is common within the governance and compliance discipline. These personnel will be asked to interpret legislation and other regulations to determine how they impact the business. They're called upon to assess the organization's current level of compliance, perform gap analysis to identify key remediation recommendations, and work with internal and external auditors to demonstrate compliance in regular reviews.

As a result of these key responsibilities, experience with legal or audit roles is particularly useful. Strong interpersonal skills are also needed, as these roles require interaction with technical teams and high-level management teams alike. Technical experience is helpful when interacting with more technical teams, but deep skill sets into specific technologies are not typically required.

2.1.7 *Education and awareness*

The human element has been a growing area of concern for some time in terms of cybersecurity. Many of the most publicized attacks take advantage of human error and social engineering at some point along their path. As a result, across all industries and sectors, educating employees, customers, users, and even the general public is becoming commonplace. The *education and awareness* discipline covers all areas of directly educating people about good security practices.

Most medium and larger organizations have implemented some form of security awareness program. The value of this training has been well studied and publicized, and many industry and government regulations require it. The makeup of these training programs can differ from one organization to the next. They can include instructor-led training, computer-based training modules, competitive activities, and internal marketing campaigns. Implementing and tracking the success of these programs typically falls on a specialized team with skill sets around education.

However, awareness training isn't the only form of education. With the massive amount of attention being paid to cybersecurity, many training programs and university degree tracks have been formed to help arm individuals with skills needed for a job in cybersecurity. The instructors who administer these types of courses are often experienced cybersecurity professionals, many of whom are still actively working as practitioners in some aspect of security.

Gabrielle Hempel is another person I spoke to in preparation for writing this book. She has a truly impressive list of experience in conducting security assessments and applying application security (particularly in terms of cloud security), and she produces educational materials on behalf of a couple of organizations. Her story is particularly impressive because she got into cybersecurity through such a nontraditional path, after working in the biomedical field. However, she recognized the links between that area of healthcare and sciences, and cybersecurity.

As I spoke to her, it became clear that she uses her varied experiences across many areas of cybersecurity and beyond to produce quality educational materials. So while she has not spent a long career in those areas, she has been able to turn that knowledge into content that effectively helps train others. This unique ability to synthesize and then share information in such an effective manner is an important focus area for cybersecurity.

As we might expect, someone with experience in education and the technical aptitude to understand cybersecurity concepts can be particularly successful in either of these types of roles. In addition to direct instruction, the ability to design a training program and curriculum is valuable in the cybersecurity world.

2.1.8 Sales and sales support

I'm sure many incredulous brows will be raised by that heading. Why would you include sales in a list of cybersecurity disciplines? Many hours were spent deliberating whether to include this. However, an undeniable fact remains about sales, sales engineers, and others who support the sales process: we wouldn't have any of those wonderful cybersecurity tools and defense mechanisms in our organizations without them.

A tenuous relationship certainly exists between IT practitioners and product vendors. However, as much as the industry might complain about some questionable tactics that come up from time to time, we need them to be a part of our community.

Selling security products doesn't just happen. Salespeople, sales engineers, solutions architects, and others all play a role in the process. They don't get there by accident either. Salespeople need to understand the concepts of cybersecurity that drive what they're selling. Sales engineers and solutions architects need deep technical knowledge of the products they're supporting in order to help customers find the right solution and configure it in a way that fits their organization.

So, *sales and sales support* are included here because these roles do help us secure the digital world. The skill sets required for these roles demand a level of understanding in security principles and even specific training on security tools and techniques. We cannot divorce the resources that help drive distribution and deployment of the defenses from the rest of the community that leverages those defenses on a daily basis.

I spoke to one such person, Carolina (Lina) Terrazas, who is a cybersecurity specialist for a networking and cybersecurity vendor called Cisco (maybe you've heard of it). She described her role as often a consultative relationship with her customers. Specifically, she meets with customers, works to understand their current environment, and makes recommendations for improving their use of existing tools or potentially investing in new tools. So while her role is a sales function, you can understand how the work she is doing has a direct impact on the overall security of those organizations.

Obviously, someone with existing sales experience and an ability to grow their cybersecurity knowledge base would find this to be a natural transition. Sales engineers and solutions architects often rely on a wide range of past experience along with specialized training on the product or products they support. However, that background experience isn't necessarily required if a person is technically inclined and can learn the product quickly along with general security skills. Of course, numerous people in more technical cybersecurity roles started in sales or sales support roles.

Terrazas's education was centered around computers and technology. Yet rather than go down an engineering path, she chose a sales-focused career. Her journey serves as an example of how technical hands-on expertise can be combined with sales principles to create real value in terms of cybersecurity.

2.1.9   *Leaders and executives*

While compiling my list of cybersecurity disciplines, I struggled with whether I should include leaders and executive-level roles as a separate discipline. To a large degree, these roles represent a culmination of the various disciplines discussed thus far in this chapter. So as you may have noticed, this is not a discipline that I included on the map I shared in figure 2.1.

However, as I talk to cybersecurity professionals about their career goals, many indicate their desire to move into a high-level role. Their common goal is to someday lead the security function of an organization as its chief information security officer (CISO). Leadership in any part of an organization requires additional skills and capabilities that aren't typically prevalent at more individual contributor-level roles. For this reason, I feel that some mention of it in this context is important.

As I said, security leaders and executives typically represent the culmination of experience within one or often multiple disciplines of cybersecurity. As the roles become more senior, these leaders need to have at least a conceptual understanding of the various functions for which they will be responsible. Additionally, they need to develop the ability to influence across other areas of the organization and communicate effectively to higher and higher levels of management.

The CISO role is often the pinnacle of a cybersecurity career. This role is responsible for all things related to cybersecurity across a large division or even the entire breadth of an organization. The CISO needs to be able to apply knowledge of cybersecurity domains as well as business management to situations that they will encounter on a daily basis. They need to see how each of these eight categories of cybersecurity disciplines fit together in a cohesive strategy.

More important however, the CISO also needs to understand how other business executives operate and how cybersecurity fits into each of their perspectives. The CISO will be called upon to update the other executives and the board of directors on the current state of cybersecurity within the company. As awareness for the criticality of cybersecurity grows in those highest levels of the organization, CISOs are increasingly engaged in high-level discussions and updates.

Security leaders in senior and executive levels in particular need to be able to communicate security in a very different way than you might expect. When speaking to other executives and senior leaders outside the security organization, technical discussions of threats, vulnerabilities, and the like may not carry much context. When attempting to gain funding for additional resources, tools, technologies, and projects, security leaders need to be capable of demonstrating how their initiatives relate to the overall success of the organization.

For this reason, while having a senior or executive-level title may seem impressive and maybe even a natural goal, these roles are not for everyone. Individuals who have a desire to understand and work in business management as much as security management will fit best in these higher-level leadership roles. They require a unique mix of technical capabilities and business administration knowledge. While demanding, these roles can also be rewarding. Having ultimate ownership and accountability for the security strategy of an organization is exciting. It is one way that a cybersecurity professional can build a legacy.

Of course, with that potential comes the risk that such a legacy may not turn out as planned. So as you look to your future, consider with a level of realistic scrutiny whether this seemingly ideal role is actually where you want your career journey to lead you.

2.2   Characteristics of a security professional

Certain skills and experiences help equip a person for specific roles within each of the cybersecurity disciplines. However, when we consider cybersecurity professionals overall, common traits apply across all the disciplines. These traits are what many hiring managers will be looking for when they're seeking security talent.

2.2.1   Inventors and creators

In relation to a lot of other fields, security is a particularly young area of focus. Additionally, it is constantly evolving. Each unique situation a practitioner encounters needs an equally unique and tailored security solution. For this reason, people who are naturally inclined to invent and create new solutions or expand upon existing technology are well suited for cybersecurity careers.

The earliest hackers often sought to understand how technology worked so they could manipulate it and create new technology based upon it. Within cybersecurity today, those principles continue to be applied. Often existing solutions and practices simply can't address the security needs of the moment. In these cases, security professionals need to be innovative and identify creative ways to address the need.

2.2.2 Obsessively inquisitive

Being obsessively inquisitive is crucial for a role in cybersecurity. When presented with a problem, security professionals need to be motived to dig in and do their own investigation. Whether it's responding to a potential security alert, assessing an application for a particular vulnerability, or trying to determine whether a company is fully compliant with the requirements of recently passed regulations, security professionals need to be deeply curious.

The answers we seek as security practitioners are not always easy to find either. So our curiosity needs to border on obsessive. Being willing to dive down the "rabbit hole" and investigate a situation until you find an answer is a common quality among the most highly respected security professionals. This persistence in working through frustrations along the way ultimately yields powerful results. This level of curiosity is found in individuals who are ready, willing, and able to learn new things. As technology grows at increasingly high rates, continuous learning is a must.

2.2.3 Compulsive learning

Not surprisingly, and related to being obsessively inquisitive, it is also imperative that aspiring security professionals demonstrate the desire and ability to learn all the time. The strongest security personnel are those who feel a level of compulsion to fully understand how something works rather than simply accept that it works in a certain way.

Understanding the inner functioning of technology enables the ability to innovate and create new solutions. This need to understand drives a person to identify portions of technology that they want to improve and to obtain the knowledge for improving it. Within the security space specifically, this applies to not only the technology systems (or humans) that we're trying to defend, but also the technology and practices that we use to defend them. Continuous improvement has shown to be effective in addressing cybersecurity at a high level. Learning and growing ensures that we're always working on making things better, safer, and more secure.

2.2.4 Idealism

Believing in and following the ideals held within the security community can be of particular value to anyone wanting to build a career in a cybersecurity-related field. Being able to see the big picture and to focus on the greater good ensures that security professionals take an ethical approach to their work.

Within cybersecurity in particular, professionals are entrusted with skills, technologies, and knowledge that can be used for good but can also be abused. Holding true to the ideals of the cybersecurity community (detailed in chapter 1) provides that ethical compass for making decisions that are consistent with the goals of this community and the organizations built around it.

2.2.5   *Forget about the "infosec rockstar"*

The infosec rockstar is a concept that was born, in part, from hacker culture and security conferences but has also further gained notoriety as social media and general mainstream awareness have brought security into the spotlight. Most people are aware that some notable personalities work within security. Hackers who discover and disclose particularly notable exploits, security professionals with highly popular blogs and podcasts, the familiar faces that are seen so often giving keynote addresses at security conferences—all of these and more are collectively referred to by some as *infosec rockstars.*

These personalities have become more visible in mainstream society as cybersecurity has become a dominant theme in the media. Bruce Schneier, Lesley Carhart, Ed Skoudis, and Katie Moussouris, for example, are just some of the influential experts that the mainstream media reaches out to for comments. Additionally, hacker culture has for decades been a mysterious element that has been glorified in movies and television. As a result, some who want to launch a cybersecurity career are seeking fame or have an unrealistic image of what working in cybersecurity is about.

Having the desire to be recognized by your peers or society for your skills and abilities can be healthy and even beneficial. But if that is your primary goal, it will likely be detrimental to your career. Analyzing your motivations for getting into cybersecurity is crucial as you start to plan the road ahead. What do you want to achieve and why? Being immersed in cybersecurity's world of constant change and uncertainty is not for everyone. Is it for you?

2.3   *Considerations of anonymity*

Within the cybersecurity community, many choose to remain largely anonymous and operate behind handles and secrecy. There are many reasons someone might want to do this. Indeed, working in security—which can include defending systems from malicious individuals or doing battle with a potentially criminal element—can make a person a target. Additionally, many of the values discussed in chapter 1 from hacker culture may cause a person to choose to protect their anonymity within the community. So why discuss this now, at this point in the book?

Well, this personal decision must be made early in your career. If you choose to protect your anonymity, that work must begin right away. While you can begin your career embracing your privacy and anonymity, and then decide later to be more present and public, you cannot go the other way. Most everyone understands that once your information is out there, it's hard to pull it back.

Although this decision may seem tactical and perhaps even frivolous, its significance could change completely a few years into your career. You should have a plan for the amount of information you want to share with the world. You need to define boundaries for yourself on the information you choose to protect. Do you want to share information about your children? Likely not. Are you OK with random individuals on the internet being able to discover where you live? Do you even want people to know your real name, gender, or other identifying characteristics about you? All of this needs to be considered.

Now this might have you thinking that anonymity is the only way to go. However, there are reasons you may choose to be openly public in your identity right away. First, maintaining that anonymity is another level of effort and stress that you will need to continually endure. How you interact with others, what you say, what you share, which websites you use—all of this and more must be considered at every moment of every day. Your level of situational awareness always needs to be high.

Second, maintaining anonymity opens you up for doxing, someone discovering and exposing who you are to the world. A certain empowerment results from not giving them that power but just being public about your identity from the start.

Third, anonymity can limit your ability to network, and networking can often be crucial in finding new jobs. You might want to build a personal brand as it opens many opportunities for you professionally. Building a brand while maintaining privacy is not impossible but is exceptionally more difficult.

No one but you can make this particular decision or tell you what is best for you. However, *before* you've launched your career is the best time to start thinking about these things and making sure that you've considered those aspects that are most important to you.

Summary

- Cybersecurity includes several high-level disciplines, and each comprises many roles and job functions.
- Highly successful professionals can be found across all of the roles and job functions within cybersecurity.
- Cybersecurity career paths encompass more than just those focused on highly technical responsibilities.
- The most successful cybersecurity professionals are inventors and creators who are obsessively inquisitive, compulsive learners, and believe in the ideology of cybersecurity culture.

Help wanted, skills in a hot market

This chapter covers

- Current state of jobs in cybersecurity and the common challenges we face
- Seniority progression within cybersecurity
- Common technical job skills employed in cybersecurity roles
- Common soft skills and how they apply to cybersecurity

In its "2019/2020 Official Annual Cybersecurity Jobs Report," Cybersecurity Ventures looked at open cybersecurity jobs. The report predicted that by 2021, more than 3.5 million cybersecurity jobs would be unfilled globally. Another report conducted by industry association (ISC)² released in 2018 estimated the number of open jobs at that time to be 2.93 million. Although the numbers may vary, every study agrees that the number of unfilled jobs in cybersecurity is a problem that's not getting better.

For those looking to begin a career in the space, this sounds like good news. A negative unemployment rate should make finding a job much easier. But not so fast. The reality is that many first-time cybersecurity job seekers struggle to find their first role. So to help better prepare you for your job search, let's discuss the current job market, some of the roles in cybersecurity, and the skills needed to land those jobs.

3.1 Job seekers vs. job openings

Analysis of the current state of the cybersecurity job market can yield confusing and seemingly contradictory results. It does seem clear that many jobs truly go unfilled each month. You can see this by simply browsing cybersecurity jobs on job posting sites like LinkedIn, Indeed, and others. Curiously, however, many people who want to get into cybersecurity careers can't seem to find jobs in this field.

What is even more interesting is that the problem is not just limited to those seeking their first cybersecurity role. Job seekers with years of experience are also having challenges trying to find that next position.

In January and February of 2020, I conducted my "Cybersecurity Careers" survey. It was actually a pair of surveys that I put out to the cybersecurity and academic communities. The first was targeted at people who had never worked in a cybersecurity role but were looking to move into the field (entry level). The second was targeted at people who were actively working or had previously worked in cybersecurity-related jobs (experienced). In both groups, those who were currently looking for a new role were asked, "How long have you been searching for a new position?" The results, shown in figure 3.1, were kind of surprising. In both groups, over 16% of the job seekers reported that they had been searching for a job for more than 12 months.

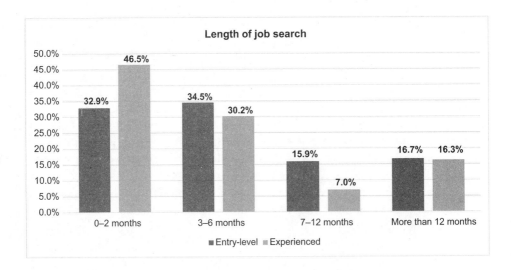

Figure 3.1   **Both entry-level and experienced cybersecurity candidates can struggle to find a new position. My survey indicates that more than 32% of entry-level candidates have been searching for seven months or longer.**

(ISC)[2]'s "Cybersecurity Workforce Study" in 2019 estimated that across the top 11 national economies in the world, 2.8 million professionals were working in cybersecurity. Within my survey, those people looking for a new role accounted for about 4.2% of the responses. That indicates that across those 11 economies, more than 27,000 experienced professionals were looking for a new job for over six months. Without a good way to estimate the number of people across the globe looking for their first role in cybersecurity, but with the growing popularity of cybersecurity majors in universities worldwide, it's safe to say the number is significant. We'll discuss later in this chapter why it's important to understand this and what this all means for someone looking to launch their career with a cybersecurity job.

Based on these numbers and what we can only speculate about the number of entry-level job seekers in the market, obviously we want to look deeper at what those job seekers bring to the table that could affect their job hunt. We might consider whether they have a degree, whether that degree is in a computer science or IT field, whether they have any security industry certifications, and even whether they've sought the help of a mentor.

These were all questions that were asked of the aspiring cybersecurity professionals responding to my 2020 "Cybersecurity Careers" survey. The results are surprising and, when considered out of context, can also be somewhat alarming. Across all of those groups, little variation occurred in the length of time people were searching for a job, as shown in figure 3.2. The shortest on average were among those who had no college degree; however, the potential for bias in this data set exists, as those numbers might

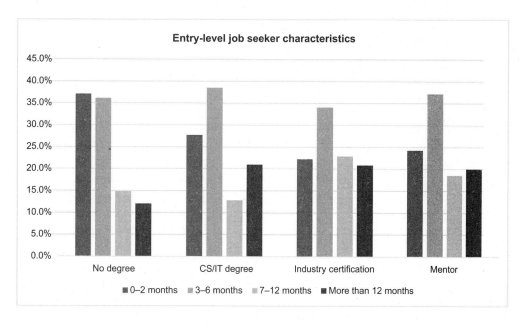

Figure 3.2 Surprisingly, survey data indicates that having a degree, an industry certification, or a mentor does not correlate to shorter job searches.

be skewed by soon-to-be graduates who are already searching for a job despite not having received a diploma yet. Still, overall the numbers are similar, which would seem to indicate that those characteristics make little difference in the difficulty of finding that first role.

For someone who has recently spent hard-earned money getting a degree or an industry certification, this may seem disheartening. Fear not; your dollars have not been wasted. A lot of factors need to be considered here for why these numbers are so similar. Additionally, those degrees, certifications, and mentors are still valuable for many reasons. We'll discuss that more in the coming chapters. For now, this data serves simply to show us the multifaceted nature of the problem and the types of challenges that a person faces in trying to launch a career in cybersecurity, and to set the backdrop for the strategies that we'll discuss throughout this book.

3.2   Cybersecurity job role progression

As discussed in chapter 2, many disciplines fit into the overall category of cybersecurity. And within those disciplines are many roles that are constantly evolving. But another aspect of cybersecurity roles is important to understand as you begin to build a plan for a cybersecurity career: *role progression*.

When we discuss role progression, we're talking about a mapping of the movement from lower-qualification roles to higher-qualification roles over time. To understand this progression, it's important to be able to recognize and identify key skills that play a part in qualifying job seekers for roles in cybersecurity.

When someone looks to start a new career path, it is common to start thinking about the big goal. Job seekers often begin to visualize where they will be after 5 or 10 or even 15 years of working in their chosen field. This is a good exercise; it is important for anyone setting out on a new career journey to have that long-term vision. Having such a vision allows for charting a course and setting shorter-term goals by which we can measure our progress, and it ultimately motivates us to keep improving and progressing as professionals.

As you begin looking at your career progression within cybersecurity, you need to understand some of the more common levels that you'll need to progress through. These roles are not specific to any discipline and instead are indicative of the breadth of responsibility that a job may hold. Understanding how careers progress through these various levels enables new professionals to identify where they want to go with their career.

This is crucial. Some people tend to assume that everyone's goal is to get to that highest level of executive management. However, for many people in technical roles, in particular, that's not the goal at all. A very different mindset and skill set are needed to do that job. Not everyone wants that set of skills to be their focus. So let's begin by looking at the various job levels and what each means in terms of responsibilities and expectations.

3.2.1 *Entry-level roles*

Entry-level jobs are the starting point for just about any career progression. These are the jobs that are your entry into the field. They should have little to no expectation of direct job-related experience and instead should focus on core skill sets that are easily adapted to the field. In cybersecurity, as we saw earlier from the survey results, it can be difficult for people wanting to start their career to find these entry-level jobs.

Organizations that are hiring for cybersecurity talent often aim to find highly experienced individuals to fill every opening. As we discuss job descriptions later in this chapter, you will see how problematic this mindset can become. Some of the entry-level jobs that do become available in cybersecurity are *associate roles*. These roles typically include a period of on-the-job training. These are great positions because they not only help the employee build their skills, but from a job-seeker perspective, also show you that the company values investing in its people.

3.2.2 *Senior roles*

As security professionals gain experience, they move up into job titles that typically are referred to as *senior-level positions*. Depending on the role, employers may require anywhere from five to eight years of prior experience for these job functions. Moving into a senior role obviously means more responsibility. Typically, these roles are expected to take the lead on specific projects or serve as mentors to other individuals within the organization.

When looking at the security job market, senior roles also tend to be some of the most common positions available. Organizations seek out experienced individuals because they are looking for people who can bring lessons learned from previous roles and additionally require minimal up-front training. In the security industry, the senior role often completes the highest volume of tactical day-to-day tasks.

3.2.3 *Architect roles*

Security architect roles are a highly specialized breed of security professional. This role requires a vast array of cybersecurity knowledge that spans multiple technologies, security disciplines, and organizational paradigms. As a result, these roles can have a base requirement of 10 years or more of previous experience. While they are some of the most highly compensated roles prior to management, they are also the scarcest in terms of open positions.

Architects are responsible for high-level security design. They need to be able to see across multiple technologies and build solutions that incorporate many defensive techniques. Architects spend a lot of time, therefore, in design work and analyzing data collected from security tools. They are expected to be able to lead teams of individuals and coach less-experienced employees as needed.

3.2.4 *Security leadership*

Moving from an individual contributor to a role that involves supervision of other professionals is a career goal or milestone that many seek. Moving into management can take on multiple forms, and the path varies from organization to organization.

Often, the first step is serving as a *team lead*. This role often doesn't carry a formal title. Instead a senior, principal, or specialist-level employee may be given the role of team lead although their actual job title does not change. In these scenarios, the leadership responsibilities given to them are typically less formal. While they are expected to provide guidance to those on the team and serve as a leader, they usually are not assigned direct reports in the HR system and typically don't have responsibility for salary management. Sometimes they may be responsible for conducting performance management tasks, but typically even in those cases the ultimate owner of performance review is the manager.

Within the security space, as with most career paths, the first HR-defined leadership title is typically *manager*. Managers have direct reports who are individual contributors. In security, a manager often has responsibility for a specialized team with a particular focus. For instance, in a large SOC, a manager may lead a team that is solely focused on monitoring firewalls, or triaging alerts from the SIEM system. However, within companies with smaller IT groups, managers may have responsibility for a multi-disciplined team that covers all security disciplines. It is important to understand this because these expectations cause a wide variance in the requirements that companies seek for these roles.

Senior managers and/or *directors* are typically the next two levels up. This role requires supervision of managers. The difference between managing individual contributors and managing other managers is significant and should not be underestimated. For this reason, companies are often methodical and conservative in filling these roles. A lot of professionals who make it to the manager role never progress beyond that because they are not well suited for that shift in leadership style. These roles also are more often filled by external candidates with previous experience than they are with internal promotions.

Responsibilities of managers tend to focus on day-to-day tactical and operational aspects of running the business. In other words, they coordinate the individual tasks of the team to achieve specific objectives. They are expected to understand, track, and report on statistics from systems and projects under their control. Often managers are given some level of salary administration responsibility as well. They also have primary ownership of hiring people into their team.

In senior manager and director roles, the focus becomes more strategic. Tracking individual tasks is less of the focus, and instead managing the longer-term objectives of the team is the primary responsibility. At these levels, planning across multiple years, measuring high-level group performance metrics, and setting a clear vision for the group is typically expected.

3.2.5 *Executive leadership*

Executive leadership typically refers to vice president (which can include associate VP, senior VP, executive VP) and officer roles such as the CISO. Many who launch a career in security see the CISO as an eventual goal. The CISO role is often looked at as the top echelon of security professionals.

However, progressing to that level requires a significant shift in focus. While VP- and CISO-level personnel need to have vast security experience, their technical background is not always prioritized as highly as their business administration skills. These executive roles are expected to understand business concepts and the way security supports and enables business objectives. Focus shifts to high-level functions like risk quantification and analysis, security program strategy, and budget planning. Technical skills play a role more in terms of being able to communicate with the security organization than on being able to execute or manage specific security tasks. This is important to understand, as not all security professionals are cut out for or would enjoy such a role.

By comparison to other executive positions, the CISO role is still quite new. It is not uncommon for technical employees who have progressed to a director level to suddenly find themselves thrust into a CISO position. Unfortunately, because organizations are still figuring out how the CISO role fits into their management structure, this can sometimes be a detrimental mistake. So it is important not to overglamorize the role of the CISO. For career success, aspirations to achieve such a level of leadership responsibility need to be tempered with objective critical self-analysis. Make sure you understand what the job really is and whether it is the right fit for you, and don't get caught up in the perceived prominence of that title.

3.3 *Common skills for security roles*

Employers typically look for a myriad of technical and nontechnical skills when they are hiring for a cybersecurity role. Understanding these skills and how they apply to security positions can help you not only plan your career development, but also better connect your skills with the requirements of a role. This is the first step in understanding the "why" behind certain requirements that often appear in job descriptions.

This section describes common skills that employers include in their job descriptions when looking to fill an open position. This is by no means an exhaustive list but should provide some guidance on where you can begin developing or improving your skills to prepare for that cybersecurity role you seek.

3.3.1 *Common technical skills*

Sometimes the term *technical skills* can be misunderstood or used in different ways. In this context, it refers to knowledge and abilities that pertain to a specific technology, whether it's hardware, software, or something else. These skills are easily measured, and, in some cases, potential employers will want to measure your proficiency in the technical skills they require. We will talk more about that in chapter 5.

With the wide range of technical skills that can apply to cybersecurity roles, it is not realistic to expect anyone to be completely proficient in all of them. Additionally, there is no one set of skills that every cybersecurity professional should have. Instead, it's important to understand how these skills are applicable to jobs you might enjoy, and then you can begin to create a plan that prioritizes developing certain skills based on the roles you seek.

SOFTWARE DEVELOPMENT/PROGRAMMING

Many job descriptions in the cybersecurity field call for some level of experience in software development or with a specific programming language. Some in the security community even go so far as to say that such skills should be mandatory. While that, in my opinion, is overstating it (such hyperbole should be avoided in any discussion of needed skills), it does hold that having knowledge of a programming language—or better still, experience in developing software—proves useful as a cybersecurity professional.

Having the ability to understand one or multiple programming languages is particularly useful since security professionals are often asked to write scripts or small programs to automate certain tasks. A strong working knowledge of scripting languages such as PowerShell, Bash, Perl, or Python is extremely helpful. PowerShell, Bash and Python, in particular, are commonly used in security jobs.

In operations roles, security professionals typically use scripts to automate tasks, search through logs, or integrate multiple monitoring systems. Scripting is useful for DFIR personnel who are responsible for helping an organization respond to security breaches and performing investigations to discover and preserve evidence that can be used later for prosecution. Writing scripts can be useful for searching through large amounts of data, launching coordinated responses, and analyzing breach evidence. Scripting even comes in handy in red-team roles such as penetration testing, where it can help conduct repetitive tasks and scan large data sets quickly.

Understanding other programming ecosystems can also be important. Most notably, helping identify and remediate vulnerabilities in software is a common task for security teams. Understanding how the code behind the software works can help when simulating attacks against software to discover its flaws, a process that the security community refers to as *offensive security*, or *red teaming*. It is also helpful in conducting static code analysis (source code review) and software composition analysis tasks. It even comes in handy to understand vulnerabilities that can be surfaced just in the way that the development ecosystems are typically used. For instance, many ecosystems make use of open source third-party dependencies. So being able to recognize when and how they're used, to identify those dependencies and the vulnerabilities they have, can be extremely helpful in application security roles.

Finally, experience in a software development role is also helpful for security professionals, especially those engaged in application security. Understanding how the SDLC and software delivery pipeline function can help security professionals better address security proactively. You will hear about *pushing left*, or executing security practices earlier in the software development pipeline. Understanding terms like *user stories* (descriptions of desired software functionality), *backlogs* (lists of user stories for future development), and *sprints* (repeating cycles of software development), and having contextual knowledge of how they apply can help you better collaborate with development teams. Much can be gained by having some level of expertise in programming and software development as a security professional.

NETWORK COMMUNICATIONS AND ADMINISTRATION

This might seem obvious, but understanding how network-connected systems communicate is extremely helpful in security roles. Many SOC analysts and DFIR employees start their careers as network administrators. The ability to understand things like TCP/IP concepts, packet-level communications, routing, and Domain Name System (DNS) pay off big when you are the one responsible for monitoring your network for attacks. Administration of firewalls and other network-based security devices definitely requires a solid understanding of network communications. Of course, if you're working in DFIR and have research SIEM data or do forensic network analysis of a recent breach, that ability to understand and interpret network communications is crucial.

From a red-teaming or penetration-testing perspective, understanding network communications is also necessary. After all, part of that role potentially requires you to manipulate traffic, detect and circumvent security controls, and so on. That would be completely impossible without the ability to analyze low-level network traffic and manipulate it to elicit the types of responses you desire or to detect anomalous behaviors.

Of course, a security professional could choose to dive into many forms of network communication. If you understand the Open Systems Interconnection (OSI) model, you are familiar with its definition of seven layers of network communications. Depending on the role you seek, you may need varying degrees of knowledge at any of those levels of the model. That said, it would be difficult to find a security professional who gained some level of knowledge in networking who can honestly say it did not help them in their job. For that reason alone, developing knowledge in this area can open a lot of doors into the security world.

CLOUD AND CLOUD-NATIVE TECHNOLOGY

More and more organizations are leveraging cloud-hosted environments—such as Amazon Web Services (AWS), Microsoft Azure, Google Cloud, and others—in which vendor-owned servers and networks replace the need for organizations to buy their own systems. These environments introduce unique demands for security and administration.

New cloud-native technologies such as containers (virtualized and often compact software-defined server modules), serverless environments, and orchestration are growing in popularity as DevOps development culture becomes more prominent. As a result, security professionals who understand the unique security threats that these technologies pose, as well as the tools used to administer these environments and technologies, are in high demand.

Becoming familiar with cloud environments and technologies can prepare you for numerous roles. Certainly, from both an operations and red-team perspective, knowing how to navigate the environments, understanding how they're configured, and being familiar with their security controls and ways to circumvent them is useful.

This knowledge can inform those in defensive roles to harden those environments better; and on the offensive side, such knowledge can help craft better attacks against the environment. Even DFIR professionals can benefit from understanding the data, logs, and tools available to them. They can use cloud-specific know-how to plan better logging and monitoring as well as be more effective in researching incidents that have occurred. As the use of cloud environments continues to grow, so will the demand for those skill sets.

CRYPTOGRAPHY

This might be another technical skill that seems obvious, but it still important to discuss. *Cryptography* is, of course, at the heart of many security controls. From protecting our communications across networks and the internet, to protecting data stored in various repositories and databases, cryptography is crucial to protecting our digital world. As a result, successful attacks against the cryptographic algorithms used to protect our data, the methods through which they are implemented, and their underlying infrastructure are the holy grail for many attackers and criminals.

Therefore, as security professionals, part of our job is ensuring that our cryptographic technologies are implemented properly, securely protected, and free from security vulnerabilities that could be exploited to gain access to sensitive information. It is important for security professionals to be able to speak to the appropriateness of specific encryption technologies as well as the risks introduced by their use.

Given how inherent cryptography is to security controls and how highly valued exploits against cryptography can be, understanding at least the basic concepts is crucial across pretty much every security role. Deep understanding of algorithms and how they are mathematically applied is likely excessive for most roles. However, being able to understand the differences between various forms of ciphers (the algorithm for protecting the data), and their characteristics that make them suitable for a specific application and not appropriate for others, is crucial.

SOCIAL ENGINEERING

Social engineering is the use of deception to manipulate people into committing an action or behavior that helps expose a secured resource. Some might argue that social engineering skills are not technical skills. My response to them would be simply, "You are wrong."

Yes, social engineering skills require an understanding of how to communicate with people. However, social engineers use specific tactics, methods, and approaches every single day. They do more than just know how to talk to people and manipulate them. They understand how the human brain works. They have a methodology you can follow to identify targets, plan an approach, and be ready to improvise on a moment's notice.

Social engineering skills also cover a wide range of functions within cybersecurity jobs. Of course, social engineering roles include professionals who are attempting to gain physical access to buildings, locations, and data via direct interaction with people. However, interactive tasks like *phishing* (social engineering via email) or *vishing* (social engineering via voice communications) campaigns might fall to someone in more of an operations role. In general, having a solid ability to influence others is an effective skill in any job, but in particular the security space, where we're often perceived negatively.

PHYSICAL SECURITY

Physical security is a concept that really starts to blur the distinction between information security and cybersecurity. Physical access controls such as door locks, electronic access authentication systems, surveillance cameras, intrusion detection systems, and physical barriers are just as necessary for protecting digital assets as cryptography, network security devices, and such.

In many organizations, physical security controls are not under the direct authority of information security teams. But under the greater umbrella of cybersecurity, it is important that those handling physical security and those handling information security work collaboratively. Many of the physical access controls employed today utilize some level of IT resources, so being able to protect those systems is crucial. These systems usually have their own unique set of threats and risks, and the types of countermeasures that can be employed to protect them can be very different from those for other IT systems.

Sometimes operational teams within the information security organization may be asked to monitor these systems and their defenses. As a result, familiarity with how they function and the unique security challenges they introduce can be useful for the SOC professional. Penetration testers might also benefit from knowledge of these systems when they encounter them on a network that is the target of their current test engagement. In addition, specific roles are available for *physical penetration testers*, those who are commissioned with attempting to breach the physical security controls of a location.

Physical security controls can be fun to learn, in part, because they are more tangibly connected to the world than more ethereal IT systems. However, it is important not to overestimate the importance of these skills in the cybersecurity context. For instance, many security conferences have lock-pick villages, where attendees can learn about the various types of locks and the tools to defeat those locks. While these villages are typically well attended, the applicability of the skills learned is minimal. Jayson Street, a well-known and highly respected physical security penetration tester, admitted in a conference session that he almost never uses his lock-picking skills on an engagement. So, while the knowledge of how locks are constructed and their weaknesses can be important at times, the ability to physically defeat them is not as crucial.

INDUSTRIAL CONTROL SYSTEMS

Industrial control systems (*ICSs*) are IT systems, devices, and networks that are used to control and monitor physical equipment and industrial processes. The equipment and processes can include manufacturing production lines, warehouse management systems, utilities management and distribution systems, or just about any other industrial system you can imagine.

ICSs present a unique set of challenges for security professionals. These systems use communications protocols and devices that are specific to ICSs; for instance, *supervisory control and data acquisition* (*SCADA*) refers to devices and communications that are used for monitoring industrial systems. Various other devices such as programmable logic controllers (PLCs), master terminal units (MTUs), and remote terminal units (RTUs), provide control of the various systems.

A particular challenge with ICS security is that many of these systems were not originally designed with the intent of connecting them to large corporate networks. Instead, their designs often assume they'll be isolated and protected from other network traffic. Adding to this challenge, many of the systems use lightweight versions of operating systems or even purpose-built firmware that isn't able to support many of the typical security features and controls that we're used to in typical IT systems.

Therefore, familiarity with these types of control systems is valuable to security professionals who work in manufacturing, utilities, logistics, and similar industries. Awareness of the need to secure these types of systems is increasing, which has begun to accelerate demand for security professionals with those skills. As a result, developing knowledge of the unique characteristics of these systems, the specialized security considerations they require, and common practices for securing them, can open doors in terms of security jobs.

RADIO COMMUNICATIONS

As connected, so-called "smart" devices and the IoT grow in popularity, so too does the use of wireless communications. This brings with it a whole new landscape of security threats. Unique security considerations apply to wireless technologies that never had to be considered when securing traditional networks.

When people think of wireless communications in terms of IT, the most commonly recognized are likely Wi-Fi and Bluetooth communications in their various forms. However, a whole variety of other specialized radio communications are leveraged in everything from smart devices to wide area networks. Even where these specialized communications pathways support a more traditional TCP/IP layer, the underlying implementations still have their own unique technology that can be exploited. As a result, each brings a unique set of risks and threats that have to be understood.

From a security professional's perspective, then, being able to sufficiently protect these communications is fully dependent upon how well those risks and threats are understood. Defensive security professionals must be able to consider questions like these:

- How and where within the stack does encryption occur, if at all?
- Can traffic be interrupted or injected, and what capabilities are designed into the communications stack to account for these issues?
- Is the underlying firmware of communications devices properly secure?

Of course, anything that is of concern to defensive security personnel, in turn, represents an area of potential attack that an offensive security professional must take into account. With organizations and individuals alike utilizing more and more wireless communications, regard for the security implications seems to be decreasing. Being able to effectively assess the security posture of these communications pathways is crucial for modern penetration testers.

It's perfectly reasonable to expect that the use of radio communications in IT will continue to increase. This growing prominence will mean that those security professionals with radio communications skills will be in high demand.

3.3.2   Soft skills

Technical skills are an important facet of finding and securing a role in the cybersecurity industry. However, professionals and aspiring professionals often over-look the importance of job-related capabilities that we call *soft skills*. These abilities aren't attributable to a specific technology or system, but rather apply across disciplines and focus, for example, on the ways people communicate and manage workloads.

Even from the perspective of a hiring manager, technical skills often receive the majority of the attention, and soft skills are given only cursory consideration or are looked at as nice-to-haves. As we will discuss in subsequent chapters, a job candidate can remind the hiring manager of the importance of soft skills for any role. For this reason, it's important for potential job seekers to not just plan for how to develop their technical skills but also make sure that they've honed their soft skills.

Soft skills can also tend to be akin to inherent talents. Certain skills just come naturally to some people, while other people have to spend time working to improve in those same areas. Being able to self-analyze and identify your strengths and weaknesses in terms of soft skills is not easy but is an immensely powerful tool in planning personal development. We will discuss some strategies for assessing your personal skills (technical and soft skills) in more detail in chapter 4. For now, let's take a look at some of the skills you'll need to evaluate.

RESEARCH SKILLS

Ask cybersecurity hiring managers what qualities they look for in a person, and two of the most common answers you will hear are *curiosity* and *passion*. The ability to look at a situation you want to understand and be able to deconstruct it and find answers is crucial to implementing effective security defenses. Some of the best security professionals are those who don't just accept a situation for what it is but instead have a desire to understand the *why*. This curiosity fuels our desire to *research* the situation and develop a deeper understanding of it. Once we understand the situation, a person's creativity and innovative spirit can drive new ideas.

It is a crucial skill to be able to start with an objective you know little about and begin looking for answers. You may not even know where to begin looking. Effective research is based in part on the ability to discover where to look and what questions to ask. Being able to answer those questions sufficiently is the end result.

To some degree, this research requires experimentation—being willing to hypothesize about what the answer may be and then testing that hypothesis. For instance, if I am seeing a certain type of alert grow in prominence in my SIEM, I might hypothesize that it is because a vulnerable device on my network is being attacked. I then might investigate the traffic generating the alerts to see if it is either destined for or coming from a specific single device. If I find out that my hypothesis is wrong, I continue to establish new hypotheses based on the information I have discovered and keep researching until I find the answer.

The applications of this skill set obviously reach far into every security discipline. Whether in a defensive or offensive security role, it's important to have the drive to fully understand the things we perceive so that we can effectively respond. Once you understand the why of the current situation, you can move on to figuring out how to start solving those problems.

PROBLEM-SOLVING

Following from that segue, the next important soft skill for security professionals is *problem-solving*. It is great to understand the situation and why it exists the way it does, but now how do you go about making it different? This is where problem-solving skills come into play. Problem-solving and research may seem like the same thing, but they are really only related concepts in that one leads to the next. Plenty of people excel at research but cannot figure out what to do with the facts once they have them. Therefore, it is important to acknowledge that the two are separate skills and to analyze your capabilities in each area.

As previously noted, problem-solving involves taking the facts that are known about the current state, identifying the desired state, and then setting forth a plan for changing the current state to the desired state. In terms of cybersecurity, this applies to so many facets and disciplines that it would be impossible to list them all here. However, a commonly encountered example is addressing the issue of user password complexity.

If I know that my users are using weak passwords that are vulnerable to attack, that is my current state. But that alone does not tell me enough to implement a solution. Next, I want to understand why my users are not using sufficiently complex passwords. Is it because the system does not support them? Is it because complex passwords are hard to remember? I need to do research to determine this. So, I investigate and find that the system does support more-complex passwords. I now know that is not the issue. Next, perhaps I conduct a user survey. I find out that people are using weak passwords because they find complex passwords hard to manage. From here, I can begin looking at solutions; I begin my problem-solving.

To solve the problem, I need to think about possible solutions and determine which will get me from the current state (in which authentication credentials are vulnerable to attack) to the desired state (in which those credentials are harder to compromise). Based on the facts, I know I need to make it easier for users to use strong credentials. I might consider deploying a password manager that would make it so users don't have to remember their passwords. Or perhaps I want to implement a biometric authentication scheme and eliminate passwords altogether. Identifying these potential solutions, selecting the one that best achieves my goal state, and then setting out the plan to do so is what problem-solving is all about.

Not all problems are as straightforward as this example, however, especially when we are talking about cybersecurity. Creativity is often needed when it comes to identifying the solutions that will bring about the desired state. The ability to analyze and come up with solutions when they may not be so obvious differentiates good problem solvers from the average person. Once we identify a solution and plan its implementation, we need to be able to work with those folks who will be affected by it and those who can help us achieve that plan to get it rolled out and put in use by the users.

COLLABORATION

Being able to work with people across the organization in a cooperative fashion is critical for cybersecurity personnel. This *collaboration* can make all the difference between being able to improve the security of our organizations and getting stuck with the same battles over and over again. Collaboration requires us to identify the people we need to work with to get things done and to work with them effectively and cooperatively.

Whether it is trying to resolve a security vulnerability found in a recent penetration test, research the increase in alerts on the SIEM, or implement a new authentication scheme, security initiatives are rarely implemented by the security team alone. Security personnel need to be able to work with other areas of the organization to get things done. If a person does not work well with others—and is abrasive, unwilling to share information, or unwilling to accept ideas from others—that person generally will not perform well in security-related roles.

Collaboration really is all about a person's ability to cooperate and compromise with those who may have different goals and responsibilities. It is the capability to influence others while also being able to accept and apply their ideas to the current situation. This is an important soft skill in many jobs within the organization. Considering how security touches every function of the company, however, it becomes all the more important in this space. If you're going to implement that new biometrics-based authentication technology, you're going to have to work with network teams, support teams, and users to make sure it is successfully deployed and used across the organization. Getting their cooperation will likely require that you're able to understand and speak to their motivations, concerns, and priorities.

EMPATHY/EMOTIONAL INTELLIGENCE

Empathy is the ability to understand and identify with someone else's emotions, feelings, and concerns. To influence others, a person needs to couple empathy with emotional intelligence. *Emotional intelligence* is a person's ability to analyze their own emotions and express them effectively. When collaborating with others, these skills ensure that a person is able to win support and cooperation from others by understanding the emotions of those they wish to work with, understanding their own emotions, and communicating in a way that bridges the gap between the two.

This type of understanding is crucial whether a person is communicating across the organization or up the management structure. Across the organization, the security professional is looking to win support and assistance from people whom they do not have authority over. When communicating up the management chain, they are trying to influence people who have authority over them. In both cases, it is important to address the concerns and priorities of the audience and motivate them to action.

This skill is undervalued by many but is terribly needed in cybersecurity roles. Many security professionals fail to progress in their careers because their communications skills are lacking. Practicing empathy and emotional intelligence addresses that issue.

MULTITASKING AND ORGANIZATIONAL SKILLS

It is probably impossible to find anyone working in a cybersecurity role who feels like they do not have a list of tasks that compete for their attention every day. Just about any security job presents competing priorities that need to be balanced, worked on in parallel, and addressed in a methodical way. The ability to take in a wide variety of inputs, organize and prioritize them, and then work through them deliberately and with purpose is *multitasking*.

Security-related job descriptions commonly reference multitasking. Unfortunately, not everyone is skilled in this regard. Some people are used to working on one task at a time, taking it from start to finish before working on the next task. For just about any cybersecurity job, this mode of operation simply does not work.

The good news is that this skill, like all the others, can be developed. Multitasking and organizational skills require a person to be attentive to details and to be methodical in planning tasks. Following a structured process can be helpful for someone who typically does not have a gift for juggling multiple work streams at the same time. So, if you find that you struggle in those situations, it would be good to look for resources on time management and organizational skills to develop your own personal method for managing your tasks.

WRITING SKILLS

Yet another soft skill that needs to be considered is *writing*. Whether you are a penetration tester, incident responder, security architect, or social engineer, you need to be able to communicate your ideas, your proposals, and your findings in written form.

Writing skills are crucial to winning support from others, whether it is for a new initiative or to fix a vulnerability. A penetration tester who cannot communicate the details of a vulnerability they discovered in a way that their customer can understand is not effective at all. An incident responder who cannot provide understandable details of a security incident to allow management to make an educated decision about a future strategy has failed in their role. And so it goes for every role in the security field.

When you are searching for a job, it is important to remember that your writing skills are one of the first things that will be assessed. You are handing that prospective employer one, possibly two, samples of your writing just through your application.

Your resume and any cover letter you choose to include are early indicators of how well you communicate in written form.

Thankfully, writing skills are among those that just take some knowledge and some practice to become proficient. Many training courses, videos, and books are available on how to become a more effective writer. If you discover that your writing skills are lacking, you should definitely prioritize developing them, especially considering the implications they could have on your job search success.

SPEAKING AND PRESENTATION SKILLS

Finally, it is also important to consider improving your *speaking* and *presentation skills.* As mentioned previously, security personnel are often called upon to communicate ideas and concepts to others outside the security space. You could be called upon to present the results of research you have performed, details of a proposal for new security technologies, or other security concepts to a group of people.

Not every role in security requires the ability to speak before large groups, but almost every role will require you to be able to effectively share your ideas and knowledge with people in nonsecurity roles. Developing skills around creating and delivering effective presentations should be a priority for anyone working or hoping to work in a security role.

Summary

- While the industry is reporting a shortage of people to fill open cybersecurity roles, both aspiring security professionals and experienced individuals can struggle to find a job. Over a quarter of those surveyed have been searching for a job for six months or longer.
- Varying levels of positions exist within cybersecurity, each introducing new challenges and responsibilities. While management is often a goal that people set for themselves in their career, it is not always the right choice.
- Aspiring security professionals should consider developing several technical skills. While no single pattern or set of skills is needed, many of the technical skills have wide-ranging applications across multiple security disciplines.
- Soft skills are commonly overlooked by people seeking their first role in cybersecurity and are also sometimes undervalued by hiring managers. Still, a person should take a look at their own skill set and work to improve the soft skills that apply across all cybersecurity roles.

Part 2

Preparing for and mastering your job search

Y ou have made it this far, which means that with a greater understanding of cybersecurity, you must still be interested in pursuing a career in this field. That is great news—congratulations!

Now that we've talked about the field, it is time for us to talk about *you*. No career journey can begin without self-examination and planning. You have seen just how vast the options are in cybersecurity careers, but you likely have questions or are unsure about which path is right for you. Additionally, you are probably beyond ready to start talking about practical strategies for launching your career. I mean, this is the *Cybersecurity Career Guide*, after all.

As you continue with part 2, we'll begin in chapter 4 by getting to know who you are and how your interests align to various roles in cybersecurity. Chapter 4 presents a progression of exercises that will assist you with this exploration. You will also learn strategies for connecting your core skills to potential jobs.

The discussion in chapter 5 focuses more on preparing your skills for a cybersecurity career path. We will examine training, academics, and certification options and their impact on your first job search. Even less formal community-based opportunities for learning are presented to give you a more comprehensive view of ways to develop your knowledge.

Finally, in chapter 6, you will find strategies for winning that first job. You will learn tactics for building the best resume you can, preparing for various forms of interviews, and even handling negotiations for the best possible job offer.

Now it is time to really dig in and start setting forth practical steps that will get you into this new career path with ease. Armed with the knowledge gained in part 2, you will be ready to start applying to jobs and working to win that all-important first cyber-security position.

Taking the less
traveled path

4

This chapter covers

- Recognizing common challenges faced by those seeking an entry-level security job
- Getting to know yourself and writing a personal objective
- Linking core skills to your objective and potential security roles
- Creating a capabilities inventory
- Identifying and documenting your skills gap

The best way to prepare yourself for your first cybersecurity job is to realize that there is no best way. My social media direct messages are filled with people who ask what seems like a simple question: "How do I start working in security?" Many are looking for a detailed step-by-step process that they should follow. Some want me to tell them what training or certification they need in order to get a job in security. The fact is, there is no one way. Perhaps more than in any other field, cybersecurity professionals have some of the most diverse backgrounds and crazy stories about getting started in security.

Deciding the correct path to take in developing your skills and positioning your-self to win that first job requires quite a bit of introspection. You have to understand what your goals are. Sure, they may change over time, but having something you are aiming for is critically important to building a career. You need to self-analyze and really figure out what motivates you, what makes you want to dig in, and what ulti-mately will spark your passion. You are going to need that motivation and passion to pursue this career.

However, if motivation and passion were enough, we would see not nearly a third of aspiring security professionals searching for over six months to find that first job. You also need to analyze your skills. In chapter 3, we talked about the common techni-cal and soft skills that are valuable to a career in this field. You need to have a process for objectively looking at where you stand today and assessing the skills you have. It is also important to understand the concept of core skills. Talents, capabilities, and experience sometimes on the surface do not have an obvious connection to cyberse-curity skills. However, when you break them down to core skills, you can then more easily apply them to a security role.

Once you understand the skills you have today and can see how they apply to potential roles that you may be interested in, your work is not finished. You now need to analyze your gaps. Look at the skills you need to have to move into that role you desire and start thinking about where you have room to further develop. This whole process is covered in the coming pages.

4.1 The entry-level challenge

Launching a new career, whether you are first coming out of a degree program or looking to make a life-altering career change, is an extremely dauting undertaking. One of the most challenging aspects of the process is just finding that first role. You need something that will allow you to grow your skills while still contributing posi-tively to the team. Within security, this seems to be a particularly difficult hill to climb. Finding a role that does not expect you to already have all the skills of a sea-soned security veteran is tough.

4.1.1 Considering cybersecurity degree programs

The education system has tried to address the shortage of cybersecurity skills. Univer-sities have more cybersecurity degree programs than ever before. Each year, thou-sands of students enroll in these programs with hopes that getting a base-level degree will improve their chances of getting that first job quickly. Unfortunately, data from my career survey shows that having a cybersecurity degree does not correlate with making the job search quicker or easier; see figure 4.1.

Do not let this data be discouraging, however. A degree is still important and can be a differentiator when competing for a job against an otherwise equally matched candidate, especially in entry-level roles. Additionally, many organizations still insist that incoming candidates have a four-year degree. Assuming that a degree was something

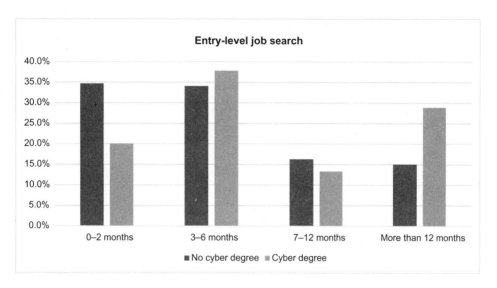

Figure 4.1 Comparing those seeking an entry-level job who have a cybersecurity degree with those who do not indicates no correlation between the degrees and improved job searches.

you were going to pursue anyway, there is no harm in having a degree in cybersecurity versus something else.

It is simply important to recognize that a degree is not a more standard road into cybersecurity than any other path. That is a critical concept that this book reinforces repeatedly: *there is no one way to get into a security career.*

4.1.2 Finding your path to security

Ask experienced security professionals how they got into cybersecurity and you'll find that many do not come from a typical computer or technical background at all. For example, Alethe Denis won the DEF CON 27 Social Engineering Capture the Flag (CTF) competition. As the winner, she now holds a lifelong DEF CON black badge, one of the highest honors for someone in the hacker community. One might expect that a person capable of winning a black badge from DEF CON must have a long security journey tracing back to their teen years or something. Well, that's not the case with Denis.

In an interview for this book, Denis talked about how she got started in security, and it was neither that long ago nor anything that one might expect. As of this writing, Denis's primary full-time job is still as a market intelligence analyst for a large staffing firm. Her entry into the security field came less than three years before her big win at DEF CON, when she started her own small business, Dragonfly Security, which she runs in addition to her full-time analyst job. She had no previous security job experience; everything she learned was either through her own investigation, self-directed learning, and networking with other professionals. And her story is just one of many.

This is something that makes the security community special but can also make entry challenging.

Why is this so important that an entire section is dedicated to highlighting the many roads that can lead to the destination of a security career? The point is, regardless of your background, you have a place in the community. Diverse sets of experience are crucial to security, as discussed previously, so you just need to learn how to identify and justify the connection between your previous job experience and the security role that you seek. Later in this chapter, we will discuss a process for doing this easily.

4.2   Know yourself

Why do you want to start a career in cybersecurity? What interests you? Why security and not a software development, project management, or marketing job? Why not be a doctor or a lawyer? What is it about security that gets you excited?

These are questions I often ask people who reach out to me via social media for help in getting their careers started. It is alarming how many of them cannot describe the motivation for their interest in security. Some tell me they see the high salaries and job security as motivating factors—maybe not the most noble of answers, but at least it is self-aware and truthful, and that is far more important. Others tell me they don't know what interests them; they just want to learn about cyber. That also concerns me, as it implies they are making a life-altering decision without a vision into what they're embarking on.

To be clear, this is not a fault, nor is it necessarily a bad thing. The cybersecurity community has done a less-than-stellar job of providing clear visibility and career mapping for starting out in security. Unfortunately, as a result, it falls to each professional to chart their own course, and that starts with understanding why you are leaving the shore in the first place. To do this, you need to take time to really analyze what it is that makes you tick and how the person you are and the skills you have make you valuable to a prospective employer.

This self-analysis will serve as the basis for building your early career. So it's important to spend dedicated time looking at what is important to you, what you want to accomplish, what assets you have currently to help make that happen, and what areas you need to focus on developing to improve your chances of success. The first step in this process is just getting to know who you are as a person and what makes you tick.

4.2.1   Identifying the authentic you

Many children are told from an early age that they can be anything they want to be and they should follow their passions. But as the weight preparing for and fulfilling adult responsibilities is loaded upon us, we are taught to be anything but true to ourselves. It is time to break out of that and get back to being the real you. When you are trying to stand out in your job application, it is exceptionally important that you embrace the things that make you unique. It seems obvious that if you want to stand out, the last thing you should try to do is craft a resume and a personal story that make you seem just like everyone else. Yet all too often that is what applicants do. Worse yet, quite often, career coaches and resume reviewers promote that behavior.

Job seekers need to be more focused on laying the foundations of a personal brand. How do you want to be seen in the industry, what do you want to be known for, or what is the story of you that you want the world to know? Everything you do in your job search can then center around that personal brand. As a first-time job seeker in a new field of work, you may not know everything about how you want to brand yourself. Just as products and corporations change their brands over time, so too might you. However, as you start putting resumes out into the hands of recruiters, you want to make sure that they are seeing what makes you awesome and special.

A colleague, Phil Gerbyshak, uses the question "What's your weird?" when he presents on personal branding. What he means is, find that one story about you that makes you stand out. It should be something fun, personal, and genuine. For him, he tells the story of being from a small town in northern Wisconsin that hardly anyone has heard of. He embraces that small-town identity to reinforce how impressive his success as a sales and personal branding trainer has been. He takes advantage of this part of his story to share how that unique perspective has shaped his approach to other aspects of his career.

This is something every job seeker, and quite honestly every professional, should do early in their career. So, take a moment and think about you. What makes you stand out? What is the crazy, odd, or just simply different aspect of who you are that few people can claim? You may not find it today; you might need some time to think about it.

But after you do figure it out, your work begins. Find that unique element and embrace it. Use it to build a story about who you are that will make employers remember your name, not just toss your information in their mental filing cabinet with all the other names they have seen. Be authentic to yourself and use that to propel yourself to the forefront of their minds.

To do that, you need to make that story a part of every aspect of your professional persona. Present it consistently in every form of autobiographical information you share. Whether it is your description of yourself on social media sites like LinkedIn or Twitter or in your cover letter that you send along with your resume, or even possibly in the objective statement you include in your resume itself, make sure you consistently share that story of who you are.

Doing so helps reinforce the connection of "your weird" to your professional life. That repetitive presentation of your distinctive persona will ensure that recruiters and hiring managers remember not just your name but who you are as a person. Making those personal connections will be key as you move into the job search process.

EXERCISE: FIND YOUR AUTHENTIC UNIQUENESS

The following exercise will help you find that characteristic that makes you authentically unique and put it into words that will guide you in building an overall personal brand for your career. It is important to set aside time to work through this in detail, and you may choose to revisit this activity multiple times. Toward the end of this section, you will use the results to help you build your personal objective statement. So, let's start finding your weird:

1. Imagine you are giving a newspaper interview in which you need to provide a brief but comprehensive autobiography.
2. List 10 to 15 things you would most want people to know about you.
3. Write one paragraph, only one, no more than 200 words, describing yourself based on that list. You may not be able to fit all the points from step 2, so prioritize.
4. Read your paragraph. List three things that you wish you had more room to describe or explain.
5. Pick the one item that stands out the most. Is it unique and does it fit you?

Now you have a possible candidate for what might be your weird. The point of this exercise is to find that thing that is inherent to the way you view yourself and is therefore authentic. However, it is also the thing with a backstory that cannot be fully described in just a few words. As a result, it makes any outside reader immediately thirsty for additional information. When it comes to your social media presence, your professional introductions, and indeed your resume, this is a priceless tool in making yourself memorable and interesting. This is what will make you stand out.

4.2.2   *Finding your passion*

Passion drives human beings to accomplish great things; it motivates us to work toward a goal despite the hardships we encounter along the way. If someone were to ask you right now what you are passionate about in security, could you immediately give them an answer?

For a lot of people, security is an exciting topic, and the idea of having a job that centers on that discipline is almost too good to be true. But, unfortunately, people who feel this way do not often explore any deeper. If you are going to try to prepare for a career in this widely diverse field, it is important to know where you want to focus your efforts. No human being can possibly learn all the skills and knowledge necessary to cover the entire breadth of security. As such, you will need to direct your efforts. Understanding what it is that motivates you and makes you want to work in this field will allow you to develop that focus.

With that in mind, take a little time to really dig into your personal feelings around topics related to security. What gets you most excited? Is it the idea of having a puzzle to solve? Is it the exposure to new technologies? Is it the constant learning and evolving? To help get you started, in the career survey, I asked aspiring security professionals which aspects of security interested them. Figure 4.2 shows the results.

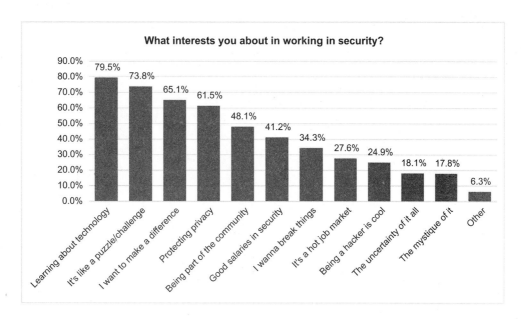

Figure 4.2   Aspiring security professionals' interests in cybersecurity

In case you are wondering how those interests hold up after someone has been working in security for a while, I also asked experienced security professionals about what keeps them interested in the field. Figure 4.3 shows those results.

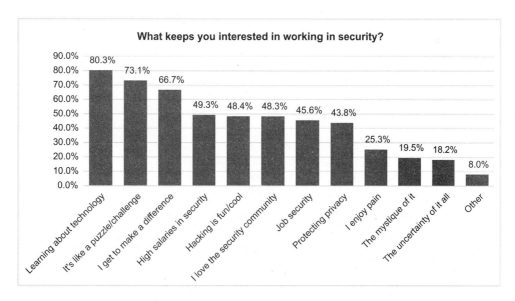

Figure 4.3   Experienced security professionals' interests in cybersecurity

You can see a lot of similarities in the responses from people who have never held a security job and those who are seasoned professionals. This is pretty telling. It's a good demonstration of how the passions that bring a person into the field continue to motivate them through their career. This is why it can be so valuable to understand your motivations in seeking a job in cybersecurity. It may not be foolproof, but a clearer view of what excites you in the field will help you focus your personal development on the key aspects that speak to your interests and help you avoid spending time doing deep research and education in areas you're just not that interested in.

Your answers, of course, may be different from the survey answers. You might have something that fits in that Other category, and that's great. Whatever those answers are for you will serve as a guide to help you find the right career path within security.

EXERCISE: IDENTIFY YOUR PASSIONS

Self-analytical activities such as this are not always easy. People sometimes find it difficult to set aside specific time to just think about what motivates them. This exercise provides a simple approach you can take to help break through that mental block:

1 Find three to five security-related blogs and/or security news websites.
2 Copy the headlines from the first five articles on each site and paste them into a document (or write them down if you prefer paper and pen).
3 Rate them in order of your interest in each article, based solely on the headline.
4 Take a look at your list and next to each headline, type or write why that article interests you.
5 Take your list of reasons and summarize them into three to five categories of interest. You might need to word those categories carefully to be inclusive. But use no more than five categories.

Now take a look at your final categories. Do those seem to match up to how you feel about security? You might want to do this on a few occasions, since news cycles and stories change daily. However, after a short time, you'll start to see trends emerge. This is a great way to establish a high degree of conscious self-awareness.

Your final list of three to five interests will be another foundational element of building out your career for the long term. They ensure that you are grounded in your goals. Now it's time to take all this work and bring it together.

4.2.3   *Developing your personal objective*

You have now spent some time getting to know who you are, how you want the world to see you, and what passions are driving your entry into the security community. You have documented the core of your personal brand and the reasons that you are seeking to build a cybersecurity career. It is time to bring that together into a personal objective statement.

Career coaches often recommend having a personal objective statement. This statement serves as a constant guide to assist you through your journey. It is something you will be able to refer to not only during your job search but also as your career progresses. You will likely encounter tough decisions in your career or moments where you question whether you are on the right career path. Having a written personal objective will help remind you of how you got to where you are and help you make a decision that is consistent with your personal best interests.

Your objective statement is personal. You may choose to keep it private or use it as part of your personal brand. At minimum, it is important to derive your personal brand from your objective statement. Maintaining that integrity of your personal brand with who you are at your core is a key to being a credible and inspirational force in your career. I always recommend that people include their objective statement, or a derivative thereof, at the top of their resume. It's that attention-grabbing sentence or two that will announce your intentions and make you stand out.

Think for a moment about the term *personal objective statement*. Let's deconstruct that in reverse. It is a *statement*; that part is easy and makes sense. It tells someone something; that is pretty simple to follow. The term *objective* indicates that it communicates something about goals or things the author hopes to accomplish. Finally, it is *personal*. This is the part that seems to get forgotten about. It should be personal to you—something that describes you, not as just another molecule in the gray mass of security professionals looking for a job, but rather what makes you, well, you. In over a decade of looking at resumes and hiring individuals for various roles, I have seen countless objective statements that were hyper focused on specific job skills or concepts but never told me about the person. Those resumes quickly become uninspiring and less than memorable.

To build an incredibly powerful statement, we are going to leverage your authentic uniqueness and your list of passions to tell a story about you and what you want to do in your career. So, if you have not done the exercises in the previous sections yet, I strongly encourage you to do so before attempting your objective statement. Let's get started.

EXERCISE: WRITE YOUR OBJECTIVE STATEMENT

Writing an objective statement simply combines your passions and "your weird" into a single statement of one or two sentences at most. It is the overall summary of who you are and why you want a career in cybersecurity. The format of your objective statement will be something like "I am (personal brand) with a passion for (top one or two passions)."

For example, my personal brand is that of a life-long hacker who saved up and bought her first computer at the age of 12. My top passion is that I love deconstructing technology to understand it and make it better. So my personal objective then becomes something to the effect of "I am a life-long hacker who since buying my first computer at age 12 has always had a passion for deconstructing technology to understand it and improve upon it." With that in mind, here's a quick process to turn your story into an objective statement:

1 What is the main idea in your personal brand? Make it as brief as possible.
2 Choose the one or two top passions from your list that you want to highlight or that best fit your story.

3 Lay out the statement as shown in the previous text.

4 Edit the statement as needed to fit your personality.

5 Ask someone you trust to read it and provide feedback. Does it make them curious to know more about you?

6 Revise if you feel necessary. (But remember, this is *your* personal statement, not the person's who provided feedback.)

Once you have your objective statement, it is good practice to reinforce it to yourself from time to time. Consider putting your statement somewhere that you will see it on a regular basis. It could be something as simple as a note on your computer monitor, something a little more artistic like a wall hanging, or something more technological like a custom login message on your phone or computer. Whatever you choose, your objective statement should inspire you and remind you of what is most important for you in terms of your career.

4.3 *Own yourself*

In the previous section, you took steps to get to know yourself better, to really understand what makes you unique and valuable, as well as what motivates you to move into a security career in the first place. Now it is time to start thinking more about yourself in the professional context. It is time to look objectively at the capabilities you have developed and the knowledge you have and to take an objective view of those characteristics that will allow you to prepare for your future in security.

Previously, I shared some of the common technical and soft skills that are often sought after by hiring managers when looking to fill a security position. Now let's look at a process for analyzing your knowledge, skills, and experience, and building an inventory of those capabilities.

In this process, you will analyze your technical and soft skills. I will share with you a process for taking seemingly unrelated technical experience from your previous roles and breaking it down to identify the core skills that can be applied across any industry or job function. As an entry-level job seeker, it will be paramount for you to be able to express to a hiring manager or recruiter just how your role as, let's say, for example, a coffee barista, has equipped you with skills that make you a more attractive candidate for the role you seek. On the surface, this may seem a bit dubious, but after reading through and trying out this process, you will gain enough confidence to present such an argument confidently and convincingly.

All of this will culminate in a ranked capabilities inventory that you will be able to use to better plan your personal development and match yourself with jobs that you are qualified for. Before getting into all that, however, it is important to understand some relevant terms.

Capabilities is ultimately the broadest all-encompassing term to understand; this includes knowledge, skills, and experience. *Capabilities* and *skill set* are synonymous in this case. Both refer to the combination of those three elements as they apply to a person's readiness for a particular job. *Knowledge, skills,* and *experience* refer to evolving levels of ability. *Knowledge* is simply an understanding of certain topics or concepts. *Skills* refers to the ability to apply that knowledge to a task or situation. *Experience*, in turn, describes a demonstrated history of using those skills.

4.3.1 Technical capabilities

Technical capabilities are probably some of the easiest to measure when it comes to self-assessment. This is the area most people think of first when they want to determine whether they are a fit for a particular job. To be fair, it is also the area that most hiring managers are focused on, so your attention on developing your technical skill set is not misplaced.

However, despite this, most job seekers rarely sit down and take inventory of their technical skills. Most people might think about it more informally and then look at a job description and see which of the boxes they can check in terms of technical requirements. A better option, however, especially for the entry-level candidate, is to have a formal inventory of the technical skills and experience you have and then apply those to a job role. Sometimes you may not have the exact skill or expertise called for, but you may have a parallel skill or expertise that would be equally useful. In the next section, I cover how to make those connections.

How do you go about identifying technical capabilities? There is no one right answer, but the crucial piece for now is to identify all possible experience and knowledge you have gained. We will talk later in this section about how to evaluate and rate each. To get started, however, it is only necessary to build a list of the skills that you have developed to any degree in your past professional and personal history.

EXERCISE: LIST TECHNICAL CAPABILITIES

Thinking in terms of knowledge, skills, and experience as they apply to technical topics, the goal here is to create a comprehensive list of those capabilities. This list will ultimately become a living document, which you'll want to update on a periodic basis or as new information becomes available.

It is not reasonable to think that in one session you will be able to sit down and list every single technical capability you have. Chances are high that you will create an initial list and then a short time later realize you need to add other items. Additionally, you will, hopefully, continue to grow and develop, causing a need to keep your list more updated. But everyone has to start somewhere, so here's the process to do so:

1 Create a document with four headings: Capability, Knowledge, Skill, and Experience.
2 Think about your current or last job first and list all tasks you had to complete in that job under the Capabilities column. Think of as many as you can but don't take more than 10 minutes.
3 Do the same for any and all previous jobs you had.
4 For all the capabilities listed, put an X in each of the three columns next to it. You did this job; therefore, you have knowledge, skills, and experience in each.

5   Next think about any formal studies you have done—either in training or schooling. List any formal studies you've taken part in that are not already listed in the Capabilities column. Again, do this quickly, no more than 10 minutes.

6   Now consider which of these you have developed skills in, and which are only knowledge, and check the appropriate columns for each.

7   Finally, think of any other nonscholarly work or self-education that you have done. List any not already listed and again check the appropriate columns. Figure 4.4 shows an example.

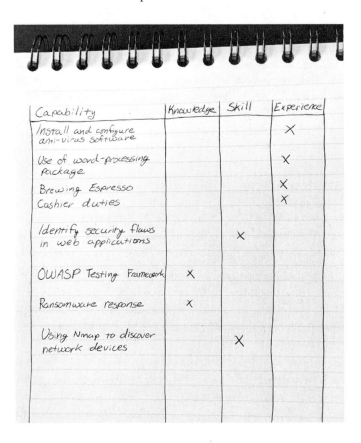

Figure 4.4   Example of a technical capabilities list

That is it. You now have a technical capabilities list. Many of them are likely not related to security, and that is to be expected. You have categorized the level of capability you have in each as well. That will be important moving forward. This will now serve as the foundation for the next step, which is identifying core skills, as well as later helping develop your capabilities inventory.

4.3.2   *Core skills*

You now have a long list of technical capabilities, many of which do not seem related to cybersecurity in any way. The next step is to connect those technical capabilities with concepts that apply to cybersecurity. For this purpose, it is time to introduce the concept of core skills.

Core skills, for purpose of this book, are the transferrable skills that each technical skill can be divided into. By understanding the core skills that are at play in each of the capabilities you identified, we will build a justification for how your past nonsecurity experience has better equipped you for that security role you are pursuing.

I made mention earlier of being a coffee barista and how someone can draw the connection of those skills to qualifications for a cybersecurity job. Examining that hypothetical situation a little deeper will give you an understanding of core skills and how to find them in your capabilities list.

Think for a minute about a coffee barista working in a busy coffee shop. What do their tasks look like? If we think about their tasks in the context of the coffee shop, it is a pretty simple list to make. They pour coffee, make espresso drinks, make other drinks, refill various supplies, clean utensils, and so forth. Sure, that makes sense, but it does not seem to apply to a security job in any way.

This is where we must identify the core skills at play. Take that list of items I just mentioned and now break it into generalized wording that does not have anything to do with making coffee. How can you describe the job of a barista in non-coffee terms? Think about what a barista actually does beyond the context of making coffee. They receive multiple inputs, possibly from various sources. From those inputs, they formulate a list of tasks that must be completed in response to those inputs. They have to arrange those tasks in a way that allows them to perform multiple tasks at the same time and maximize their efficiency. Additionally, they have to plan and execute mission-critical maintenance activities (refilling, restocking, cleaning) while creating minimal impact to the efficient delivery to the customer.

Did you ever stop to think about it that way before? When stated in these terms, the core skills that could be applied to a cybersecurity role become more apparent. For instance, think about someone in an SOC analyst role. Take the preceding description and apply it to common tasks that a SOC analyst might have to complete on a daily basis. Is there anything in that description that doesn't fit?

- Processing multiple inputs
- Translating inputs into response tasks
- Prioritizing and organizing responses to maximize efficiency
- Planning and executing critical maintenance activities
- Maintaining focus on efficient delivery for the customer

How about a different role in security? Think for a moment how this same description of skills would apply to another role. Maybe try applying it to a penetration tester position, or incident response, or even a security salesperson. Now you start to see how core skills can be identified and used as justification for a person's qualifications for a job in a completely different line of work.

When you're approaching an entry-level job and don't have a lot of past experience in the field to draw on, being able to confidently justify these connections will prove valuable. For now, let's focus on the process for identifying the core skills and we can discuss later how to then apply those core skills during the application and interview process for a particular cybersecurity job.

EXERCISE: IDENTIFY CORE SKILLS

For this exercise, you need the capabilities list that you made in the previous section. This process may seem obvious, given the previous barista example, but I will lay it out step-by-step here for easier reference:

1 From your technical capabilities list, take all the nonsecurity-focused capabilities that you marked in the Experience column and put them in a new list.

2 For each capability, reword your description, replacing any terminology that would identify the job or industry to which it applies or allow someone to narrow it to a subset of industries. Remember, the goal is to generalize it enough that it can be applied universally.

3 After you have completed rewording the descriptions, break them into singular skills as shown previously in the barista-to-SOC-analyst example.

4 Look through the singular skills and eliminate any duplicates (you likely will have quite a few).

5 Test out each skill to see if you can apply it to two or three careers from multiple other fields. If you cannot, try rewording the bullet further to truly capture a job-neutral description of the skill. Figure 4.5 shows an example core skills list.

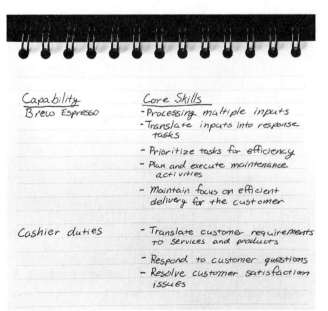

Figure 4.5 Example core skills list

This exercise may seem a bit oversimplified, but the results are invaluable when you're in a conversation trying to justify why you are the right candidate for a job even though you do not have the experience of a veteran professional. If you were able to do all these steps in your head, that is great. However, for most people this is a learned skill that takes time and practice. Writing it down and going through the process step-by-step can assist in making it a more reflexive ability.

The danger in trying to deliver this type of justification is that if it is not delivered well, it can come off as desperation or exaggeration. Therefore, once you have your core skills list completed, it is crucial to practice describing those core skills as they would relate to a hypothetical job. Confidence drives credibility in this case. If you can deliver the message with conviction and passion, the receiver of that information is far more likely to accept it.

Finally, don't make the mistake of confusing core skills with soft skills. The two concepts are different. Core skills will ultimately apply to a specific task or capability of a potential job. Soft skills remain a little more ambiguous in that sense. They are not related to any one specific task and, instead, apply across the myriad of responsibilities for a given job role. Therefore, soft skills are the final list we need to look at before working on building a capabilities inventory.

4.3.3 Soft skills

As discussed previously, *soft skills* are those abilities that do not apply to a specific task, technology, or system. Instead they are typically associated with higher-level personality traits and help augment a person's ability to complete many job-related responsibilities more effectively. Soft skills are a critical element when hiring managers are looking at potential candidates. For instance, the most highly skilled penetration tester would still be useless as a security consultant, a job that relies heavily on the person's ability to convey information to the customer, if they have poor communications skills.

Unfortunately, soft skills are hard to self-analyze. It takes an incredible degree of self-awareness and objectivity to see where we excel and to admit where we are weak. No matter how objective we try to be, a level of personal bias always remains simply because we cannot perceive ourselves the way the outside world perceives us. Measurement of soft skills tends to be more subjective.

Objective measures that attempt to assess soft skills generally do not sufficiently quantify the full scope of the person's ability. For instance, typing speed and grammar can be used to objectively assess a person's writing ability. However, that is not sufficient. A person can write a grammatically perfect, 30-page report in one day and score extremely high in those measures. However, if the report fails to create understanding of the information with the readers, is it really a success?

Such is the nature of soft skills, and I regret to tell you that I have not found the oracle of knowledge that can provide a solution for this. We must accept that no matter how hard we try, measuring our soft skills is still prone to error, and the measure of our capabilities lies ultimately in the minds of those around us. Still, common practices and perceptions lead to relatively similar assessment of people's soft skills overall.

For this reason, it is still valuable to look at your own abilities so that you can ultimately determine which are your strengths and which are areas you should prioritize for further development.

Finally, most soft skills are things that you can practice improving. However, a certain level of innate capability can also be a deciding factor in how much you will have to work on them before you can demonstrate certain skills. At times you may simply have to be aware of weaknesses that you cannot improve upon and compensate for those in other ways.

All right, soft skills are tough to measure. Then how do you go about determining if you are able to apply those skills and if you have demonstrated them in the past? Well, since beauty is in the eye of the beholder, that beholder might be the right place to start. In other words, subjectivity means that every person's perception and values (along with other predisposed characteristics) will determine how they rate a given quality. Therefore, if you want to know how well you perform in a subjective matter, you need to sample those who have been in a position to judge you on those qualities.

I'm not suggesting that you go out and survey people you have worked with in the past, although that would be one effective way to discover your strengths and weaknesses. Instead, you've likely already been in positions where you've received such feedback. If you have ever worked in a job before, you have likely gotten some type of performance appraisal from your manager. Even if it was just informal commentary, it would be enough for you to gather some necessary information. If you have never been employed before, you still have likely gotten feedback through your schooling, interactions with friends and family, and possibly interactions with other people.

Do not forget that feedback can be both positive and negative and is not always direct. People will not always tell you, "Oh, that was a really empathetic way that you spoke to me about that problem." Instead they give us other indicators, so you are going to need to think back on situations where something worked or did not work as a result of a soft skill you attempted to employ. So, let's get to building that list shall we?

EXERCISE: LIST YOUR SOFT SKILLS

This process is much easier to describe. While identifying soft skills accurately is hard, the process also has a certain simplicity, given that we are relying heavily on outside individuals to help us discover our strengths and weaknesses. Follow these steps:

1 Beginning with your current or most recent job, think about the soft skills feedback you have received, both positive and negative. Take no more than five minutes to list the skills for which you have received feedback. Refer to chapter 3 if you need a reminder of common soft skills.

2 Put a plus sign next to the ones that were positive feedback, and a negative sign next to the ones that were areas you were told you need to improve. If you end up with more negative than positive, that is OK, and you don't need to feel bad. Studies have shown that humans tend to notice and remember negative feedback more easily than positive.

3 Next, or if you do not have employment history to use, think about what your friends and family have told you in the past. Are there certain characteristics about you that your friends have noticed in a positive or negative way? List those the same as before.

4 Think about any other interactions you have had where people commented on any of your soft skills and list those as described previously.

That is it for now. You have now developed three comprehensive lists that help describe your capabilities that will affect your preparedness for a cybersecurity role. But something is still missing: we need to rate your proficiency with each and determine where you need further development and which capabilities you can tout as your greatest strengths.

4.3.4 The capabilities inventory

All of the exercises in this section have been leading to one thing: the creation of your *capabilities inventory*. As you might suspect, your inventory will have three sections that correspond to the three lists that you have created thus far. The trick now, though, is that you are going to use all that you have learned through this process to provide relative ratings for your proficiency in each. This is where the proverbial rubber meets the road because this inventory is going to guide you as you work on developing your skills to align with the job you want, which we will cover in the next section.

Having a solid capabilities inventory comes in handy beyond your job search. Building such an inventory is a good objective way to take a step back and assess where you are at. Few professionals take the time to do such an inventory, and as a result they tend to make assumptions about their own abilities that may not be true. Additionally, they fail to recognize where they have room to improve, and that can lead to frustration or even stagnation (as they simply stop learning new skills).

I encourage you to keep these materials you have created thus far and continue to update them as your career progresses. You may choose to remove items that become clearly irrelevant. As you develop more security-related skills and gain more experience, the focus on core skills will become less, and you may at some point choose to stop including those altogether. The point is this can be a living document that will serve you for years to come. It is a lot easier to maintain once you have created it rather than going back and trying to re-create it 5 or 10 years down the road.

EXERCISE: CREATE A SECURITY-RELATED CAPABILITIES INVENTORY

Gather up your technical capabilities list, your core skills list, and your soft skills list. Then do the following:

1 Since the focus of your inventory is on finding a security job, take only the security-related capabilities that you captured in your capabilities list (the others are captured as core skills, remember).

2 Replace each X in the Experience column with the number of years of experience you have in that capability.

3 Replace each X in the Skill column with a rating of your proficiency in applying that skill. Use a four-level scale of Beginner, Advanced, Comprehensive, or Expert.

4 Replace each X in the Knowledge column with a rating of your familiarity with that topic, using the same four-level scale.

5 Look at your core skills. Each of these came from a job responsibility you had at some point; therefore, you have experience. Enter the number of years of experience you have for each, remembering to combine the experience of multiple jobs where appropriate.

6 Considering your core skills, rate them in order from greatest to least proficiency. You want to rate them relative to one another, as this will help you as you begin to apply them to security roles.

7 Add in your soft skills list. For those marked as positive, rate how commonly you receive that feedback. Use a three-level scale of Rarely, Common, and Often. Do the same for those marked as negatives. Figure 4.6 shows an example completed capabilities inventory.

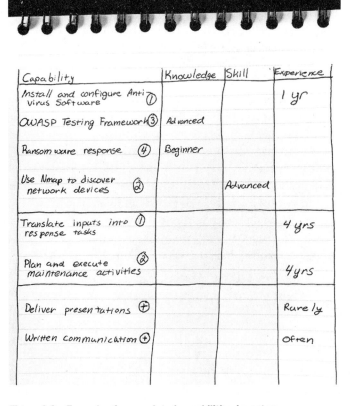

Figure 4.6   Example of a completed capabilities inventory

Wonderful, you now have a detailed capabilities inventory that will help you in selecting jobs you're qualified for. This list will help you justify your qualifications to potential recruiters and hiring managers, and plot out a plan for further developing your skills in preparation for getting a job.

4.4 Connect yourself

So far you've gotten to know yourself by taking a look at what makes you who you are and what passions are motivating you to find a job in the cybersecurity space, and you've taken ownership of yourself by examining your capabilities and building out your capabilities inventory. Now it is time to connect yourself. It is time to move from understanding and owning yourself, to aligning yourself with the cybersecurity job that you want to have. All the work you have done has led you to this point.

The goal of connecting yourself is ultimately to use your objective statement and your capabilities inventory to identify where you fit in terms of potential cybersecurity jobs and where you will want to prioritize further development to help you be better prepared and qualified for a job.

4.4.1 Choosing a focus discipline

You built a personal objective statement that describes what makes you unique and what excites you about cybersecurity. Now it is time to take that information and use it to figure out which roles in cybersecurity are most likely to fit your passions and keep you interested for a long-term career. Your objective statement details your core values that you need to stay true to above all else. Compromising on your objectives and your passions will only lead to frustration down the road. After putting in all the effort to start a career in cybersecurity, the last thing you want to do is go down a path that conflicts with your passions and causes you to seek a whole new career outside of security.

Depending on how aware you already were of your passions, you may already have a discipline or two in mind. That is great if that is the case. However, your driving curiosity may also lead you to have strong interests in a lot of areas of security. This is why we put together your list of passions the way we did, to help you focus.

Now it is time for you to start doing some research. If you already have an idea of the type of role that you want to get into in security, take a look at the common roles I detailed in chapter 2. Look at the responsibilities of those jobs. Additionally, do some searching online and find others' descriptions of what the jobs you are interested in entail. Now, compare those responsibilities to the passions you rated at the top of your list for your objective statement; do they align? If so, great, you are on the right path and have chosen one or a couple of potential disciplines that should be a good fit for you.

If they do not align or you are not sure which role you want to get into, start from the opposite angle. Start with your objective statement and the passions that you listed as most important to you. Now reread the cybersecurity discipline descriptions in chapter 2. Research online to find out more about the responsibilities of different roles in security and see which ones seem to align with your passions. Consider reaching out if you have a network of people on social media and ask others what they feel might be a good fit for someone with your passions.

Ultimately, seek out the roles with the right responsibilities to keep you interested in the long run. You want your career to cultivate your passions, because that is what will lead to longevity in this field. Ultimately, come up with one or two disciplines you want to focus on for your career.

PIVOTING

Picking the discipline that will define the rest of your career in security may seem like a lot of pressure. But if it turns out you have picked a discipline that doesn't fit your passions that way you expected or if your passions change down the road (as they sometimes do), you are not out of luck. Within security, specializations that I have identified as disciplines do exist. However, common concepts apply across those disciplines, and some of the specialized knowledge or skills from one discipline can provide unique context that is valuable in another discipline. For this reason, security is a profession in which pivoting from one role to another is much easier than in other industries.

With that in mind, do not get too bogged down in trying to find the single right role for you. Trying to find the right path is important, but those paths can change, and new paths arise as new technologies and needs develop across the digital world. If you are feeling anxiety or fear over choosing the right direction, try to let go of that and understand that your path will be unique and likely unexpected, just like so many others in this industry.

4.4.2 *Identifying gaps and challenges*

Well this is it; this is the moment you have probably been anticipating through much of this chapter. You have completely inventoried yourself. You have analyzed which career paths in security might be the best fit for you. Now it is time to look at how well you are prepared today for a job in your chosen discipline and where you will need to further develop your skills to better align yourself with the responsibilities and requirements of the job you're looking for.

The process here is pretty simple. It will require you to have a solid understanding of the job that you seek and the typical requirements for that job. To do so, you will have to put in additional research effort. The good news is this can be kind of fun.

EXERCISE: DOCUMENT YOUR CAPABILITIES GAP

Your *capabilities gap* is simply the list of capabilities that you do not currently have or do not have a sufficient degree of to match up with the job you are seeking. You know from the capabilities inventory you created where you excel, but you need to know what hiring managers are looking for in those roles you are want to pursue in order to make a comparison. So, get your typing fingers ready; you are going on a job hunt:

1 Compile 10 to 15 job descriptions for jobs in your chosen discipline. You want to search job boards for the lowest level of role (entry level, if possible, or the most junior level of experience).

2 From those job descriptions, compile a unique list of the requirements listed for each. Count how many times you see a particular requirement show up in those descriptions.

3 For those requirements that state a certain length of experience, document the average number of years across all appearances of that requirement. Ultimately, we're trying to get a good selection of data and normalize it to some degree.

4 Once you have the list, select the five most commonly seen requirements.

5 Do the same for additional or preferred qualifications; list them uniquely, counting the number of times they appear and then select the top five.

6 Compare the list of five required and five preferred qualifications to your capabilities inventory.

7 Start with your technical capabilities: do any of them match up? Do you have the number of years of experience? Start a list and include any that you already fulfill completely, marking them as Complete Match. Next, if you have any that match in capability but not in experience, list those as Need Experience.

8 Take any of the top fives from Required and Preferred that you did not match to a technical capability and compare them to your core skills. For any that match, add those to your list and indicate that they are a Core Skills match.

9 For any of the requirements that did not make it to your list so far, check whether they are soft skills. If they are, do they match your positives? If so, list them and label them as Positive Soft Skill on your list. If they are negative, list them as Negative Soft Skill.

10 Add any remaining requirements or preferred qualifications that did not match anything from your capabilities inventory to your list and label them as Not Matched.

You now have your capabilities gap documented. Take a look and think about prioritization. Here is how you'll want to prioritize these as you start thinking about your personal development plan (this is covered in further detail in chapter 5):

- *Not Matched*—These list items are your highest priority. These are likely technical capabilities that are popular with employers that you do not have a match for. We will discuss strategies in the next chapter for how to begin addressing these needs.

- *Core Skills*—These items represent an opportunity to further develop a direct technical capability. However, you at least have a credible way to demonstrate value to fill this qualification, so this is next on your prioritization list.

- *Negative Soft Skills*—These are areas you need to address, but as far as your immediate job search, you can move them down in terms of priority because they take time to develop. One caveat: if any of them are soft skills you listed as hearing about often, you may want to consider raising the priority a bit.

- *Need Experience*—There's nothing you can do to develop demonstrated job experience. You can prioritize these items for working on your own to better apply these skills, but that should not be a priority if other items have higher priority.
- *Positive Soft Skills*—Good news: this is a strength you can highlight on your resume and in your conversations. Keep practicing these skills to get better, but you don't need to prioritize working on them.
- *Complete Match*—Good news again: you completely fill this qualification. Keep your capabilities fresh, but you don't need to prioritize these for development.

Once you have completed this prioritization, you are ready to begin building out a plan to address your personal development. You know exactly where you want to focus your skills development to get the job you are looking for. The next chapter presents various ways to go about attempting to address these capabilities gaps and develop at least the necessary skills to justify yourself as ready for the job.

Summary

- Aspiring security professionals face key challenges when trying to land their first job. Entry-level roles are hard to find, job descriptions are often unrealistic, and no single path is defined for starting a career in security.
- A personal objective statement ties together a unique aspect of your personal brand along with the elements that you feel most passionate about in your chosen career path. It can serve as a constant guide for making career decisions.
- Core skills are an important way to demonstrate how nonsecurity-related work history and experience apply as qualifications for a security job.
- A capabilities inventory is crucial for understanding how well prepared you are for a job in your selected security discipline.
- Documenting your skills gap based on real-world job descriptions provides a powerful planning tool for personal capability development, allowing job seekers to better align with the jobs they seek.

Addressing your
capabilities gap

5

This chapter covers

- Identifying cybersecurity certifications, assessing their value, and choosing whether to pursue them
- Understanding how academic cybersecurity programs impact the job search
- Leveraging less formal and self-directed education to build practical skills

---

For as long as I have been mentoring people looking to start careers in security, the most common questions I receive are about deciding what training or certifications they should be pursuing. When people are getting ready to launch their careers, they inevitably think about the technical skills required to land their first job. In the preceding chapter, you learned how to self-analyze, choose the right path for your security career, and identify the gaps in your capabilities. Now that you have a better view of the technical capabilities you need to focus on to prepare for a career in security, it is time to examine ways to fill those gaps.

I mentioned earlier that the security community has not done a good job of providing a clear pathway for people who are looking to launch their career. You will receive different answers about the best place to start, depending on who you ask.

Unfortunately, I think many responses come from a place of wanting to be helpful but not actually having good information.

For years, security leaders speculated that the problem with finding security talent was that the educational system was not covering cybersecurity in their curriculums. As a result, we have seen cybersecurity topics covered in curriculums all the way down to the elementary level. Colleges and universities have launched full-scale degree programs focused on cybersecurity, in some cases even creating specialized programs that focus on specific areas of security. However, because of the nature of such degree programs, graduates are sometimes told that the knowledge they received was too general or out-of-date.

In response, some experienced and well-meaning mentors began telling aspiring security professionals that they should get a certification, which would equip them with the skills they needed to get started. So, people started getting security certifications. When it didn't work, they got more certifications. Pretty soon the industry was being inundated with entry-level candidates who had a laundry list of certifications but no demonstrated experience. And thus, security professionals started telling them that certifications were not meaningful without experience and that they needed to have more real-world experience applying security concepts.

To address this, more recently the prominent advice from seasoned professionals has changed to suggesting people take part in learning activities. Often, they'll recommend CTF competitions, where individuals or teams attempt to exploit vulnerabilities to find hidden flags that earn points. Hackathons are another common suggestion. These events typically feature the full group working collaboratively on a security initiative. Others will often suggest that new learners build their own virtual labs to experiment with security technologies and offensive security techniques.

Unfortunately, in a corporate setting where job descriptions have requirements that include minimum years of experience, recruiters look for verifiable experience. It can be difficult to present participation in these more informal educational activities in a credible fashion that can fulfill those requirements. As a result, those who just want to get started with a career in a field of work that has a supposed skill shortage are left wondering where to go next.

This conundrum of "I need experience to get a job, but I need a job to get experience" has been a decades-long obstacle for hiring security talent. Unfortunately, despite the efforts of many in the security community to improve this situation, the increasing attention and scrutiny of companies following high-profile breaches has only served to make it worse.

This chapter is focused on helping you overcome those challenges from the job seeker perspective. Specifically, we will focus on the skills and experiences that bring the most value as an entry-level candidate and how you can best demonstrate those skills to prospective employers.

5.1   *The alphabet soup of security certifications*

If you have looked into a career in cybersecurity, you've probably quickly become aware of the myriad of *certifications*. These designations, created by third-party organizations as a way for professionals to demonstrate proficiency in a particular skill or set of skills, are almost ubiquitous to cybersecurity job descriptions. As a result, for many aspiring security professionals, getting one, two, or even ten certifications can seem like a great way to prepare for a job in cybersecurity. However, as we saw from the results of the "Cybersecurity Careers" survey in chapter 3, holding an industry certification does not correlate to a shorter job search.

Despite the survey results, there are very good reasons for having a security certification when searching for a role, especially your first role. The most compelling reason is simply that most employers ask for, and many require, an applicant to have a security certification of some type. It might seem simple then, that an aspiring security professional should dive in and work to achieve a security certification. However, it is not that simple.

Just the sheer number of available options can be difficult to decipher and understand. Choosing a certification is not just a matter of picking the one that is most commonly asked for in job descriptions. Many certifications have requirements for previous knowledge, work experience, or prerequisite certifications. The chart in figure 5.1 can give you some context for just how complex the security certification landscape has become.

Domain	Organization	Beginner	Advanced
General knowledge	CompTIA	Security+	CASP+
	SANS Institute	GSEC	GISDP
	(ISC)^2	SSCP	CISSP
Penetration testing	CompTIA	Pentest+	
	Offensive Security		OSCP
			OSWP
			OSWA
	SANS Institute	GPEN	GXPN
		GWAPT	GMOB
			GCPN
			GAWN
GRC & management	ISACA	CSXP	CISA
			CISM
			CRISC
Cloud security	CompTIA	Cloud+	
	SANS Institute	GCSA	GCPN
	(ISC)^2		CCSP
Application security	(ISC)^2		CSSLP
	SANS Institute	GWEB	GDAT
			GFACT

Figure 5.1   This chart depicts many common security certifications and their relationships to career progressions.

Choosing a certification is more difficult than it might sound, yet it can be an important step in finding that first security role. For that reason, it makes sense to look at some of the most common security certifications and understand where they fit in an overall security career.

5.1.1 Security certification overview

Before digging deeper into specific security certifications, let's take a look at the purpose behind these certifications and how we came to have so many of them. Cybersecurity (and indeed, its predecessor, information security) were not always the well-recognized disciplines that they are today. Many of the formal education and degree programs available now simply didn't exist. Even now, such programs often struggle to keep up with the pace of change and the wide breadth of subject matter included under the cybersecurity umbrella.

With this lack of formal education, organizations like (ISC)², ISACA, and others sought to provide ways to formalize the knowledge that was being developed by information security professionals and share it across the community. Eventually, they would go a step further and create certification programs enabling professionals to demonstrate their mastery of the subject matter by taking an exam. A passing score resulted in the person being recognized as a certified professional.

As cybersecurity has evolved, more and more specialized knowledge areas have emerged. To help recognize professionals who have demonstrated expertise in those areas, additional cybersecurity certifications have been launched. Additional organizations such as the Computing Technology Industry Association (CompTIA), SANS, the International Council of E-Commerce Consultants (EC-Council), and Offensive Security have established additional certifications that focus on areas of general knowledge as well as specialized topics. Certifications provide a common level of measurement and an attestation of a professional's skills. As such, they became a tool used by recruiters and hiring managers to help locate and evaluate potential candidates.

In addition to these security community organizations (many of which operate as nonprofits with the stated goal of improving security education), commercial security vendors have created certification programs specific to their products. For instance, Cisco Systems has a wide variety of certifications that demonstrate proficiency in not only its products, but how to use those products in the context of specific disciplines such as security. Some of these are shown in figure 5.1.

As cybersecurity has grown, so has the demand for these types of certifications. Along with the certifications, educational programs and content designed to help professionals prepare for the certification exams have expanded as well. Many of the materials and programs are provided by the certifying organizations themselves. However, many third-party educational organizations have built their businesses on helping people prepare for these certifications. The materials, courses, and programs that lead to a security certification have become increasingly focused on not just preparation but actually training in the needed skills to pass the exam. The subtle shift from preparation to training has caused many to view certifications as an educational process, not simply an attestation of skills.

The result of this shift is that many current and aspiring professionals use certification programs as their primary way to learn new skills. Right or wrong, this means that certification holders may not have demonstrated skills in that area outside of a classroom or lab setting. For much of the industry, this has been viewed as a negative, a detraction to the perceived validity of the certification itself. To combat this, some of the certifying organizations have implemented work experience requirements. Many of the certifications include a requirement for demonstrating continuing education on an annual basis. Some certifications even have a progression; one certification must be achieved before a professional can then complete the next higher-level certification.

All of this, of course, impacts an aspiring professional's potential choices in certifications. It is not uncommon to encounter job descriptions that call for a certification that cannot be achieved without prior working experience. This makes the role appear out of reach for potential entry-level candidates even when, in some cases, the role is an entry-level job. We'll discuss job descriptions and how to recognize and overcome these issues in chapter 6. For now, let's look at some of the most common certifications you'll see specified in job listings.

5.1.2 (ISC)² Certified Information Systems Security Professional

One of the most recognized security certifications, in part because it's also one of the oldest, is the *Certified Information Systems Security Professional* (*CISSP*), administered by (ISC)² and launched in 1994. This certification is one of the most commonly listed in job descriptions. In a sampling of job descriptions from five major job search websites, I discovered over 90% specifically mentioned the CISSP as a requirement or preferred qualification. Some directly stated that the CISSP was required, but the majority included phrasing to the effect of "CISSP certification or equivalent."

So, what is the CISSP certification? This highly respected, general-knowledge certification covers topics that span eight disciplines, or domains, of knowledge. These domains include the following:

- Security and Risk Management
- Asset Security
- Security Architecture and Engineering
- Communication and Network Security
- Identity and Access Management (IAM)
- Security Assessment and Testing
- Security Operations
- Software Development Security

As you can see, the CISSP certification covers some pretty broad topics. As a result, this general certification can apply to just about any role in cybersecurity. However, the complexity of a CISSP certification extends beyond the wide breadth of knowledge required to pass the exam. To obtain a CISSP certification, the applicant also needs to demonstrate five years of experience working in at least two of the domains. This is meant to ensure that certified professionals have expertise in applying the subject matter to real-world situations and precludes simply taking a bootcamp-style prep course to gain all the necessary knowledge.

(ISC)^2 goes to great effort to ensure the quality of certification holders. Beyond the qualifications needed to achieve the certification in the first place, (ISC)^2 also requires CISSP holders to demonstrate continued learning through the continuing professional education (CPE) program. CISSP-certified individuals are required to complete a minimum of 120 hours of continuing education over the course of three years. They are required to register their CPEs with (ISC)^2, which can perform audits to ensure that all CPEs were properly earned through approved educational activities.

Finally, let's talk about the costs for a moment. Certifications are not free. There is a fee for taking the exam, another fee for the certification itself, and of course if you take any form of preparatory course or purchase any preparatory materials, those all have a cost associated with them too. As of the writing of this book, the exam fee for the CISSP is $699. That's no small fee, especially for someone who's looking to launch their career. The annual maintenance fee to keep the certification current is $85. Preparatory courses for the CISSP can vary greatly from one vendor to the next but can cost as much as $3,000. That's a lot of money to spend up front before you're even working in a security role.

If this sounds daunting to you as an aspiring security professional, that is probably as it should be. The CISSP is meant to be a somewhat prestigious designation of established security professionals. As you can see from the requirements to obtain and maintain the certification, it is not appropriate for someone just entering the industry. As you're seeing all these jobs posted that list a CISSP as a requirement, understand that this is not a route you can take early on. Don't be disheartened, however, since as you'll notice, most of those job descriptions also include the phrase "or equivalent," and that's what we'll focus on as we prep you for starting your career journey.

Now that you've seen that the CISSP is not a realistic route for a new or aspiring professional, let's talk about another way to demonstrate that you hold a security certification to help you check that box on the job description you're looking at.

5.1.3   *CompTIA Security+*

A certification that has been around for quite a while but seems to receive less attention is CompTIA's *Security+* certification. A bit newer than the CISSP, the Security+ was formally launched in 2002. The CompTIA website (www.comptia.org) identifies six main topic areas that are included in the Security+ certification:

- Threats, Attacks, and Vulnerabilities
- Technologies and Tools
- Architecture and Design

- Identity and Access Management
- Risk Management
- Cryptography and PKI

Although they're arranged differently, these topics overlap considerably with the CISSP domains. The goal of the CompTIA is to be general in terms of certifying the skills of a wide range of security professionals, much like the CISSP. CompTIA provides the following description of the latest version of the exam on its website:

> *The CompTIA Security+ certification exam will verify the successful candidate has the knowledge and skills required to assess the security posture of an enterprise environment and recommend and implement appropriate security solutions; monitor and secure hybrid environments, including cloud, mobile, and IoT; operate with an awareness of applicable laws and policies, including principles of governance, risk, and compliance; identify, analyze, and respond to security events and incidents.*

The Security+ is a bit more attainable for an aspiring security professional than the CISSP. While CompTIA recommends that candidates pass their Network+ certification and have at least two years of work experience in a security field, it is not a requirement. A person who has gained the necessary technical knowledge through other means (education or self-study, for example) can still take the exam and obtain a Security+ certification. The Security+ does come with requirements for continuing education. A Security+ holder must complete 50 continuing education units (CEUs) over a three-year period to keep their certification.

Financially, the Security+ is more accessible as well. The exam costs, at the time of this writing, $379. That's still not cheap, but over $300 less expensive than the CISSP. Annual renewal fees for the Security+ are $50. While preparatory courses are offered by many vendors, in general they are cheaper than the CISSP as well. E-learning courses are available through CompTIA directly for prices ranging from $499–$899. Instructor-led courses cost more, but even the most expensive package is about $2,000.

Overall, the Security+ assesses a broad range of security topics and knowledge, which makes it applicable to just about any security role someone would want to pursue. In that sense, it is very much like the CISSP. However, the costs and requirements to achieve a Security+ make it an appropriate first certification for someone looking to land their first security job and to bring a certification to the table. While certain job descriptions—in particular, government jobs for many countries—require the CISSP specifically, most are open to other certifications that demonstrate the candidate has a measurable level of expertise in cybersecurity topics and technologies.

The Security+ is growing in popularity among people looking to launch a career in security, but it is not the only certification available. Other options exist as well, although some can be a bit more specialized in terms of the types of roles that they would apply to. Next, we'll take a look at one such certification.

TA 492 1980

5.1.4   *EC-Council Certified Ethical Hacker*

Initially launched in 2010, EC-Council's *Certified Ethical Hacker* (*CEH*) certification is a relatively newer option. As you may have guessed from the name, it focuses on skills for the ethical hacker (typically referred to as penetration testers). The CEH certification has undergone many revisions, both in terms of its content and the programs used to administer it. Many of these revisions have likely come in response to criticisms from the security community regarding the rigor of the certification process. As such, while this certification is acceptable to most employers as a demonstration of fundamental knowledge in penetration testing, it is not as well respected overall as other available certifications.

From its inception, the CEH was intended to provide more specialized training and assessment of penetration testing tools, techniques, and concepts. As you may have noticed, I said *training* and *assessment*. Unlike many of the other certifications available, including those we've previously discussed, the CEH more closely ties the training to the exam itself. While it is still possible to take the exam without going through a formal training course, EC-Council requires you to have been working in security for two years. It requires a verification and application process that includes sending in your resume along with a $100 Eligibility Check Payment in order to purchase just the exam. It is also notable that EC-Council is one of the few certification organizations that operates as a for-profit commercial business rather than a non-profit organization.

According to the EC-Council CEH Exam Blueprint version 3.0, the exam assesses knowledge across seven domains:

- Background
- Analysis/Assessment
- Security
- Tools/Systems/Programs
- Procedures/Methodology
- Regulation/Policy
- Ethics

The CEH certification courses are open to anyone without formal prerequisites. While EC-Council does provide suggestions on the types of people who should enroll for CEH training, there are no enforced requirements. This makes it an accessible certification for aspiring professionals. Like most other certifications, the CEH has a requirement for continuing education. To maintain a CEH certification, EC-Council requires 120 EC-Council continuing education (ECEs) credits over a three-year period.

Costs for the CEH are a bit different from others we've looked at. Given the difficulty in trying to register for an exam without taking the training, it is also then difficult to determine what the fee would be. EC-Council does require certification holders to pay an annual membership fee of $80 to keep their certifications current.

What is different here is that it is a membership fee for the organization, so one fee covers all EC-Council certifications the person may hold (they do offer several other more advanced and/or more specialized certifications). Training for the CEH ranges from a self-paced E-learning course for $1,899 as of this writing to $2,999 for an in-person training course.

Overall, the CEH, while accessible to aspiring security professionals, does have significant drawbacks. Cost is a key issue unless you already work for a company that is willing to help pay for the certification so you can transition to a security role within the company. Another drawback is the particularly focused nature of this certification. If you know for sure that you want to work in penetration testing, the CEH might be a good investment. However, it could potentially be limiting in terms of your ability to pivot to other areas of security if you find that penetration testing simply is not for you.

5.1.5 Other certifications

The certifications discussed are some of the most often talked about for aspiring security professionals. However, as you saw in figure 5.1, many others are offered by a variety of organizations and product vendors. By better understanding the CISSP, Security+, and CEH, hopefully you now have a clearer idea of the important aspects that need to be considered when you are looking at certifications now or in the future. Each certification has its own scope of knowledge it assesses, its own requirements for continuing education, and costs associated with it. Some of the more technical certifications even require a practical exam that requires you to complete tasks in a lab environment within a given time frame.

Discussing each certification in detail is beyond the scope of this book. An entire book of its own could probably be written on the topic. However, the landscape of certifications changes so fast that it's more important to be able to analyze and assess the value of each in terms of your own goals than to know the intimate details of every certification available. However, one more question needs to be answered in terms of certifications.

5.1.6 How many is too many?

The previous section title, "The alphabet soup of certifications," is a reference to a term that many in the security community use to describe the long list of certification acronyms that some people place after their names on resumes, business cards, and so forth. Indeed, many people working in security today have amassed a long list of certifications, sometimes 10, 15, or even 20.

It can seem like more equals better, but is that really true? Is there a point of diminishing returns? What about the complexities of trying to manage and meet the renewal needs of each of those certifications? What about the costs? These are all things you'll need to consider as you think about the number of certifications you want to pursue. But much of that is fodder for a point deeper into your career. Right now, we need to look at what you need to get started.

I admit, I have yet to see a job description specify that a person needs to have more than one certification. While, yes, sometimes having a few key certifications can make you stand out or augment your capabilities for a particular role, in general it's not required. The key to having a certification to get you started is simply to get you past that bar of just having a certification at all. Many employers require it, others at least prefer it, and if nothing else, being certified shows that you take your skills development seriously and have enough expertise to pass the exam.

Now that is not meant to be a declaration that a single certification is all you should get before you land your first job. Other factors need to be considered. How would an additional certification benefit you in the job search? If you have a Security+, would getting a CEH or a SANS GIAC Penetration Tester (GPEN) make you a more attractive candidate for that associate penetration tester job? Perhaps it would. However, a cost is also associated with obtaining and maintaining those additional certifications. Is it worth it? Only you can decide. Maybe you are considering enrolling in a program that will allow you to achieve a Network+ and a Security+ certification for cheaper than if you did them separately. At that point, perhaps it would be more worthwhile to do them together. These are decisions you will have to make as you progress.

Taking things to the extreme, too many certifications can be a hindrance to you in the job search. Especially as a newcomer, having many certifications can give the hiring manager the impression that you have put more effort into obtaining certifications than in actually learning and applying the knowledge. This is a real phenomenon that I have encountered myself when interviewing entry-level candidates. While I personally would never exclude a candidate from consideration just based on having a lot of certifications, my experience has shown that usually once we talk, I find that they've failed to fully grasp the knowledge that was supposed to have been gained to pass those exams. It's a case of studying and learning to pass the test but then failing to comprehend and apply the knowledge to practical real-world situations.

Returning to the goal at hand, which is to get you into that first cybersecurity job, obtaining that first certification has significant value. Beyond that, the return on your investment starts to decline quickly. A second, more specialized certification might help you stand out for a specific set of roles, so it may be a good idea. Adding more certifications on top of that is a costly endeavor that won't likely impact your goal and could ultimately hurt you in the job search process.

5.2   *Academic cybersecurity programs*

As was discussed earlier, a couple of decades ago very little existed in terms of formal education offerings for information security/cybersecurity. However, that landscape is changing significantly as cybersecurity has become not only a popular topic on the daily news but also a popular career path for many individuals.

Colleges and universities have been adding degree programs and even advanced research labs that focus on cybersecurity topics. Students can enroll in cybersecurity degree programs ranging from associate degrees to PhD-level work. Research programs and labs are also available to students to further expand their skills and experience.

This seems like a no-brainer right? Get a cybersecurity degree and get a cybersecurity job. Well, not so fast. Let's break this down and look at just how necessary and effective academic programs are in kicking off a cybersecurity career.

5.2.1 Degree programs

As cybersecurity has grown into its own career path, the business world has put more and more pressure on academics to provide formal preparation for careers in this space. Many institutions have launched cybersecurity degree programs. Even more impressive is that many high schools are now starting to introduce courses in cybersecurity as well and/or integrating cybersecurity topics into Computer Science Advanced Placement (AP) exams.

With all the attention being paid to cybersecurity degree programs, a person wishing to launch a career in security may feel compelled to enroll in one of these academic majors. However, as part of the cybersecurity careers survey, experienced cybersecurity professionals were asked about their degrees. Figure 5.2 shows the results.

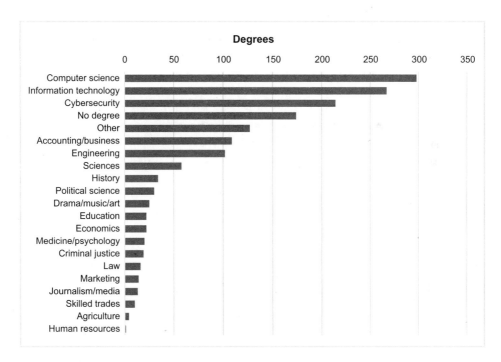

Figure 5.2   Degrees held by experienced cybersecurity professionals

As you can see from the data, computer-related degrees definitely dominate the space. That is to be expected. In particular, it is notable that the third most common degree is in cybersecurity. This shows the value that such degrees can have. However, does this mean that if you already have a degree in a different field or have no degree at all that you should enroll in a cybersecurity degree program tomorrow? Absolutely not. Looking again at figure 5.2, you can see that the fourth most common response was no degree at all! From there down, you can see significant numbers of responses across a wide range of degree subjects. No doubt, many of these responses came from individuals who did not start off in a security job and instead pivoted from a job in a completely unrelated field.

In preparation for writing this book, I interviewed experienced professionals in cybersecurity. I introduced you to one of them, Alethe Denis, in chapter 4. One thing that I found, perhaps not too surprisingly, was that many of the people I interviewed did not start in computer-related fields at all. However, as their roles began to touch on areas that were focused more on security aspects, they found they had an interest in the field and made a pivot in their career. Anecdotally, I can share that it is common to hear security professionals talk about how they got into this field by accident. I was initially enrolled as a pre-med major in college. It was only after a few semesters that I transitioned to a computer science degree and began working as a software developer. I worked in that programmer role for almost a decade before I finally transitioned over to a security role.

As you saw in chapter 1, security touches on just about every facet of our lives. As a result, it is easy to pivot into a security career from another field not directly related to technology at all. So back to degree programs, what, if any, degree program should an aspiring professional choose? As with so many things, the answer is, it depends.

As noted earlier, a significant number of security professionals do not hold a degree at all. However, this does not mean that it is advisable for an aspiring security professional to forego a college degree altogether. While many organizations are moving away from hard requirements for four-year degrees, most still seek either a degree or comparable work experience in a related field. Indeed, most of the experienced professionals who do not have a degree worked in some form of technology field for years before they moved into security. Therefore, for an aspiring security professional who does not have previous work experience in a related field, landing a job without some form of degree would be difficult.

OK, so a degree is necessary; does that mean it has to be a degree in cybersecurity? Here the answer is an unequivocal no. First of all, for those looking to land their first role who already have a degree, going back to school for a degree in cybersecurity would have little impact on their job qualifications. Considering the cost and time of university degree programs, there are better ways to spend those resources that would have far greater impact on job readiness. For those entering college who want to eventually move into the security workforce, a cybersecurity degree can be more helpful. Essentially, you're going to school anyway, so getting a degree in your intended field makes sense. However, it is not required.

Looking again at survey results in figure 5.2, you will notice that the most common degrees are computer science and information technology degrees. To be fair, those numbers are certainly biased since many experienced professionals entered this field before cybersecurity degrees were even available. However, even today, employers are typically open to various degrees when looking for security resources. In fact, the curriculums of computer science and information technology majors cover a lot of related subject matter that can be useful for a security professional. Understanding networking, software development, operating systems, and other topics covered can be invaluable to a security professional.

What should the aspiring professional do? Just as we talked about in chapter 3 for choosing a job pursuit, it is really most important that you choose a degree program that best fits your interests and passions. This will help ensure that you remain motivated and engaged throughout your studies, which ultimately results in an outcome when you enter the job market. So, take some time to really read through the course catalogue. Understand the material covered in each degree program and pick a program that covers topics you find interesting and exciting. There are other ways to develop additional security-related skills that will make you an attractive candidate for prospective employers.

5.2.2   Advanced research programs

While choosing the right degree program is important, so too is choosing the right university or college. It is important to look beyond just the degree itself and instead really understand the supporting activities and programs that the school offers. Some of the more valuable programs starting to gain popularity are external research groups and labs. These programs provide students an opportunity to build practical skill sets by performing real-world security research.

An example of one such program is the University of California at Santa Barbara's Computer Security Group, known simply as SecLab. The SecLab team devotes its time to various initiatives that center around "designing, building, and validating secure software systems." While this may sound theoretical, the work they do has practical real-world impact. For instance, part of their work is analyzing open source software for security weaknesses. When they find weaknesses, they report them not only to the maintainer of the software but also to security organizations that ultimately report the findings to the industry at large. In this one aspect alone, SecLab is helping make software more secure.

Other programs focus on collaborative partnerships with industry organizations working on various projects and research. For example, MITRE is an organization that manages various federally funded research centers in the United States. It regularly partners with universities and other academic facilities to support its research. This gains MITRE access to researchers, while the students involved have the opportunity to build practical real-world skills that are useful when they begin searching for their first cybersecurity role.

When looking for a school and a degree program, it is important to understand the types of opportunities available. Especially when pursuing a noncybersecurity-related degree, the ability to still participate in this kind of research provides excellent material to make a resume stand out. However, if the school you choose does not offer these kinds of opportunities, all hope is not lost. You can gain practical cybersecurity skills in other, less formal ways that you can ultimately include on your resume.

5.3   *Less formal skills building*

As a result of many forces, from information sharing to high demand for skilled workers, cybersecurity offers many opportunities for learning. While these are not formal degree programs, they can often provide more contemporary and practical skills that you simply would not gain from an academic degree program.

A significant challenge with these types of learning opportunities is reflecting them on a resume in a way that is credible and substantiates your skills. Let's examine the more common, yet less formal, methods of learning and developing additional cybersecurity skills.

5.3.1   *Industry conferences*

The security community has a long history of hosting conferences to provide opportunities for information sharing, skills development, and professional social networking. Typically, these conferences feature presentations from skilled individuals from across the community sharing their ideas, results of their latest research, detailed analysis of technologies, or other novel security-related topics.

Additionally, these conferences often offer the opportunity to attend formal training or workshops as part of the event. Many also host *villages*, hands-on learning centers dedicated to a specific technology or technique. For instance, a popular village seen at many conferences is the lock-picking village, as noted in chapter 3. Participants can learn about the construction of physical locks by learning to pick them from experienced professionals working in the village. The focus of villages can range from car hacking to ICSs to voting systems and more.

Numerous types of conferences exist. Understanding the target audience and the subject matter of the conference will help when choosing which ones are of most interest and most value. Conferences include those focused on security practitioners, specialty conferences, and even nonsecurity conferences that can all be great learning opportunities.

Security practitioner conferences tend to focus on people who are working or desire to work in an organization helping to build and maintain security defenses. The subject matter of these conferences can be wide-ranging across multiple security domains. The focus is typically on activities that would be leveraged inside an organization to manage the security preparedness of that organization. Some examples of common security conferences that focus on security practitioners include the RSA Conference, Black Hat conferences, and InfoSec World. With the wide-ranging topic areas covered at these conferences, attendees can choose to attend trainings and speaker sessions that fit the areas they want to focus on most.

Specialty conferences also are hosted throughout the year. These conferences tend to focus on a specific discipline or domain of security. For instance, one of the most popular and longest-running conferences in the world is the DEF CON hacker conference. This conference began in 1993 as an informal meetup of hackers in Las Vegas. Since that time, it's grown to be one of the largest security conferences in the world. Its primary focus is on the hacker community and hacker-related topics. However, as it has grown, so has the community of villages that surround the conference, providing wider breadth of subject matter. Other well-known hacker conferences focus on topics of offensive security, including the BSides series, SchmooCon, THOTCON, and others.

Shifting away from hacker cons, another well-known specialty conference is the Layer 8 conference. *Layer 8* refers to humans as a component of security. In other words, the Layer 8 conference focuses on topics related to social engineering and intelligence gathering. While many of the related activities like villages and workshops provide a wider range of topics, the primary focus of the conference itself is that human element and tactics for attacking and also securing humans.

Yet another specialty conference is the OWASP Global AppSec conference. OWASP is an industry organization dedicated to application security, securing software, and software development processes. Again, while this is the primary focus of the conference, other activities and even some of the speaker sessions may overlap other aspects or domains of security.

An entirely separate book could be written on all the various forms of conferences and each individual event. It is not important nor particularly reasonable to be familiar with every topic area and every conference available. However, it is important to understand how these conferences can be useful for developing knowledge and skills that will be helpful to you on your journey into a security career.

First, simply attending one of these conferences is a learning opportunity you can list on your resume to demonstrate knowledge in a particular domain or across various domains. Your attendance shows not only the knowledge you have gained, but also your desire to learn and grow as a security professional. That can be an important consideration for potential hiring managers.

Second, taking advantage of the workshops, training, and village activities can give you more practical skills that you can list on your resume. Being able to show that you took time to not only gain knowledge but also practically apply knowledge and develop skills in a specific area can be highlighted when more formal job experience is lacking.

Third, and even more significant, the majority of these conferences host a call for papers (CFP) in which anyone can submit a topic that they would like to share as a speaker. If you have done research into a particular domain or topic in security and have a novel idea or fresh data to present to others, that is perfect material for a conference presentation. Even presenting on a more common topic but sharing it in a different way than has previously been done can be valuable, especially at smaller community-focused conferences.

Regardless, being able to share on your resume that you spoke at a conference (or multiple conferences) is particularly eye-catching for a prospective employer. To be fair, for most aspiring professionals, this might be a high bar to overcome. However, if you feel comfortable with public speaking and have material to share either from your own independent research or as part of group research, it is not an impossible task either.

One challenge with conferences is that they are not free. They cost money for admission (unless you're a speaker or can find a scholarship ticket). Additionally, they're located all over the country, so unless you happen to have one that is hosted local to you, you'll likely need to travel, which incurs its own expenses.

One solution is to take a look at the BSides series of conferences. These independent conferences, sanctioned by the BSides organization, are all a little different, but most seem to have a wide breadth of topics. Additionally, hundreds of them occur in major cities around the globe, so the chance of finding a local one is much higher. Many of them are smaller community-sized conferences, and therefore getting accepted to speak can be easier as well. In the end, conferences are a great way to gain some knowledge, and maybe some skills, but you do have to consider the cost involved.

5.3.2 *CTFs, playgrounds, and personal labs*

You can gain additional knowledge and skill in security in more informal ways as well. Sometimes these are a part of a conference; other times they exist as stand-alone events. Capture-the-flag (CTF) events are common both at conferences and as special events sponsored by various industry organizations. In these competitions, individuals or teams attempt to complete security challenges in order to retrieve "flags." The security challenges can be focused on offensive security "hacking" practices, social engineering, OSINT, or some even include defensive components. Different CTFs will provide differing levels of guidance as far as the challenges, with the most extreme providing only the target environment that the user must explore with no instruction at all.

These events can range from beginner-friendly to extremely competitive. Beginner-friendly CTFs generally focus more on the teaching and guiding aspects and deemphasize the competition side. They generally provide specific detailed challenges and hints for completing them. In many cases, those administering the CTF will also be willing to help if you get stuck on a challenge.

Conversely, some CTFs, especially those at larger conferences like DEFCON, can be exceptionally competitive. However, do not count out those competitions completely, especially if they are team based. If you can find a team whose members are willing to take on a new person looking to learn, they can still be of tremendous benefit. Either way, participation in CTFs should definitely be something you include as resume material. I will talk more about this in chapter 6.

Playground labs are another terrific way to gain experience in security practices. *Hack The Box* (*HTB*) is an example of a ready-made playground in which users access and follow challenges to hack various servers. New servers are made available with new challenges on a regular basis. Scores are kept and users ranked. A crucial differentiator for HTB is that employers list open job postings on the site, and they are matched to you based on your ranking. HTB is still growing, but if you are interested in penetration testing and vulnerability research, it's a great way to demonstrate your skills with credibility for potential employers.

Other options are available for playground environments as well. One of the longer-standing options is an intentionally vulnerable application from OWASP called Web-Goat. You can set up the application on your own computer and practice different forms of attacks. While this is great for learning pentesting skills, understanding the various types of attacks can also be useful in defensive roles. Various other intentionally vulnerable playground environments exist that you can work with as well. Some are available over the web; many require you to set up an environment to run it on your personal computer.

That brings us to personal lab environments. *Virtualization*, the ability to set up virtual computers that run as software on your personal computer, allows for setting up small network environments easily (and more important, cost effectively). Building a small network in a virtual environment on your home computer can allow you to experiment and do research with lots of networking and security technologies and practices. It's a great way to expand your knowledge, especially if you're the type of person who likes to tinker with things. Even just installing and building the environment itself will teach you valuable skills. With Google, YouTube, and other easily searchable media on the web, you shouldn't have a problem locating helpful information when you get stuck.

5.3.3   *Webinars, podcasts, and live-stream events*

The last category of less formal learning I will talk about are online media such as webinars, podcasts, and live streams. Webinars are often offered for free by vendors of security products or even industry organizations. Podcasts are almost always free to listen to and are usually presented by prominent members of the security industry. Live-stream events and videos are growing in popularity, with many people providing free how-tos and learning workshops.

Attending, watching, or listening to these types of media is not typically something you will look to include on your resume. However, they are valuable for multiple reasons. First, they help keep you informed of current trends, topics, technologies, and debates that exist in the security space. This can help you gain knowledge that will help you have more informed conversations in the interview process. It will also give you ideas for further exploration and learning on your own.

Second, these types of media will expose you to a wide range of personalities within the security industry. You will hear differing ideas on how to solve problems.

You will begin to absorb the common vernacular used by security professionals, which again can help you have a stronger conversation in the job interview process and make you stand out as a candidate.

Third, and maybe this should be the most obvious, you will potentially learn new skills from these sessions. Vendor webinars are often set up to introduce product capabilities or demonstrate how to accomplish certain tasks with their products. Livestream events are often in the form of how-to or demonstration videos from people working in security. Even podcasts can sometimes have an educational component as guests discuss their work or things they have discovered through research. Ultimately, these are free resources that can guide you and help you form more knowledgeable opinions about accomplishing security-related objectives.

5.3.4   *Other community meetups*

A final resource for learning new skills is local security community meetups. Numerous national and international organizations have chapters in various cities. These chapters typically have meetings or other less formal meetups where the members can get together to discuss security topics. Some of these events have speakers delivering formal speeches on various security-related topics. Others might have workshops or interactive activities for teaching skills. Still others may be even less formal gatherings for drinks or meals, meant to foster interaction and networking among the members of the group.

All of these can serve as valuable opportunities to gain knowledge and possibly practice new skills. The following is a list of some of the national and international communities with local chapters that gather regularly:

- Cloud Security Alliance (CSA)
- DEF CON Groups
- Information Systems Security Association (ISSA)
- InfraGard
- International Association for Healthcare Security and Safety (IAHSS)
- ISACA
- (ISC)^2
- Open Web Application Security Project (OWASP)
- Women in Cybersecurity (WiCyS)
- Women of Security (WoSEC)

Summary

- The cybersecurity industry has a vast array of certifications across various domains. It is important to find the right balance of cost versus quality.
- Having many certifications does not make it easier to find a job in security. One or two certifications will get you past requirements on job postings, and that's all that an aspiring security professional should be looking to do early on.

- Cybersecurity degrees are not required to get a job in security. What is more important is having a degree in something you are passionate about.
- Various institutions have other cybersecurity programs that provide the opportunity to develop practical skills in real-life settings.
- Conferences, workshops, CTFs, labs and even online media can all be powerful learning tools that help build resume content or increase your cybersecurity knowledge.

Resumes, applications, and interviews

This chapter covers

- Building and tailoring a resume to accommodate applicant tracking systems
- Choosing the right job opportunities to apply to
- Making informal learning and skills development stand out credibly
- Preparing for screening, technical, and team interviews
- Avoiding common pitfalls that often trip up candidates
- Negotiating and accepting terms of a job offer

So far, we have taken a look at the cybersecurity landscape, the roles that fall within it, and the skills required for those roles. We have talked about you, how to find your interests, and how to align your skills to the job you want. Many obstacles stand in your way of landing that first job, and we have acknowledged those as well. So now it's time to take all that prep work and put it into action. Let's dig into the key aspects of the job search process itself and talk strategies for being successful and avoiding common pitfalls.

Three main steps make up the job search: building your resume, choosing and applying to jobs, and going through the interview process. We are going to go through each step and talk about how to put yourself in the best possible position to be successful. We will also look at common mistakes applicants make and ways to avoid them. As you might expect, we will start by talking about your resume.

6.1 Mastering the resume

What is a resume? Many people would say it's a document that lists details of skills, work experience, and education. All of that is true. However, I would like you to think about your resume in a different way: think about it as your portfolio.

Your resume makes your first impression with any prospective employer, before they ever have a chance to speak to you. It is the summation of all the work you have done, and it potentially includes more than just that document you're going to send along with your application. When you think about your resume as a portfolio rather than simply a document, it leads you down a creative path. Being creative in your resume allows you to express yourself, which conveys your authenticity. That is what ultimately will make you stand out as an applicant.

6.1.1 One document is not enough

I have talked to many aspiring security professionals who tell me they have sent their resume to large numbers of potential employers. It is the same story every time. They spent a bunch of time creating the perfect resume, following all the advice they could dig up. They worded, reworded, revised, edited, and tweaked it until they had the perfectly formatted and articulated document. Then they sent it out to lots of employers who had jobs they were interested in and wondered why they get so few calls back.

Their mistakes begin with that one document. For as long as I can remember, I have read advice on how to write a resume. Almost without fail, each article, blog, and column has made the assumption that job seekers create a singular one-size-fits-all resume. I guess it makes sense, right? You are talking about yourself, so of course you want to do that in the perfect way. But that could not be more wrong. You should have multiple versions of your resume. Let's look at why and how you need to do this.

Your resume does indeed introduce you. But it is not an autobiography. As I mentioned previously, your resume is the first impression that an employer will get of you when you apply to a job. However, they are not looking to hear interesting stories about a person and all the wonderful things that person has done. When prospective employers are looking at resumes, they are trying to understand how the person they are reading about is going to bring value to their company. So, your resume is your first chance to tell the story of how you are going to bring value to them in the role that you are applying for.

On an episode of my podcast, my cohosts and I had the honor of talking with Jake Williams, a well-respected member of the security community. One of the things that he shared with us is a strategy he uses when coaching folks on building their resume. It goes something like this. He starts off by having them share their *elevator pitch*. They get one to two minutes to convince him why they are the right fit and how they are going to bring value to his team in the role that they are applying for. At the end of the two minutes, he asks them to point out where in their resume he can find the topics that they just shared. Anything that is not there is something that needs to be added. Anything that is in the resume that they did not include in their pitch is something they might think about removing.

I really like this exercise because it stresses the idea of your resume being that first elevator pitch. It is your opportunity, in a brief window, to win over the person who might consider you to fill their open position. However, one aspect that's missing is critical here: the best elevator pitch is always tailored to the audience you are speaking to. Your resume needs to do the same.

Each job you are applying to is unique. Even if they have the same title, call for the same skills, and require the same level of experience, they are still different. Each company and even each team has its own challenges it is looking to solve by hiring a new person. Your focus as a job applicant needs to be on understanding those challenges and crafting your resume to show that you are the right person to help address them.

It may sound crazy, but you should have a separate version of your resume for every single job you apply for. On the surface, that seems like a lot of work. But often you may find that you have a couple of versions based on the kind of role you might apply to, and then only minor tweaks are needed to tailor it for the job at hand. You can keep each version, and as you find similar openings elsewhere, just work from what you did previously. With that in mind, let's analyze key facets you need to consider.

6.1.2 Format

Job applicants do, to some extent, understand that their resume is their first chance to shine, their first chance to stand out. Certain "experts" offer advice that you want to have a resume format that is fresh and different, something that is pleasing to the eyes. This can be good advice for the humans who are going to read your resume. Adding in a light dose of graphics, some color, multiple columns, maybe even a good headshot of yourself can personalize your resume. That is all great, except it misses one key aspect of the modern job hunt: *applicant tracking systems* (*ATSs*).

Many organizations today use these automated ATSs to post and receive applications for their open jobs. You can easily recognize them because many times they are the job sites that ask you to upload a resume and then attempt to parse it and fill in all your details (with varying success). But this step is crucial because those systems are filling in all the data that will initially be used to understand who you are and how you are qualified. The better job the ATS does in processing your resume, the better results you can expect.

For this reason, when you think about format, you need to think about simplicity. How can you make your resume easy for the ATS to read it and properly process it? You should think about a few factors here:

- *Overall format*—Stick to a single column, rather than trying to get fancy with multiple columns.
- *Graphics*—Skip graphics, as they will add nothing for the ATS and only make the document harder for the automated parser to understand.
- *Fonts*—Use a font that is easy to read and common such as Arial, Times New Roman, or Calibri.
- *Simple header*—Include a simple header with your basic personal contact information.
- *Section headings*—Use clearly worded and obviously formatted section headings.
- *Standard sections*—Include standard sections such as Education, Skills, Work Experience, and Certifications.

This is great, but do you still want to have a creative resume to provide to humans? That is fine. Keep two copies, one with all the fun, individualized formatting that you can send in advance of each interview, and one that you will use whenever the application is being entered into an ATS.

One last word about format. People commonly ask how long their resume should be. This is a matter of opinion, and I have seen a wide range of answers. As a hiring manager, I will share my view. Anything longer than a two-page resume, for me, becomes onerous and boring to read. Chances are, if you are including so much employment history or other background information that you need more than two pages, a lot of irrelevant information is probably included. This is a good time to examine the information you are including and how it applies to the objective of selling yourself as the best candidate for the job.

6.1.3 Check the boxes

One of the biggest challenges when applying to any job posting is getting your resume past the ATS to be seen by the recruiter and hopefully the hiring manager. In the previous section, I shared simple strategies for using formatting to help the ATS understand and process your information accurately. However, I also shared how important it is that you tailor your resume to the position you are applying to. This is probably one of the most effective tools to pushing past the ATS and even the recruiter and ensuring that the hiring manager gets your resume in their inbox.

I mentioned earlier that the ATS is responsible for parsing the information on your resume and filling in those important details in your job application. However, you need to be aware of another even more crucial function when dealing with an ATS. Each application processed by the ATS is given a score. The ATS compares your resume information with the requirements listed for the job you have applied to. From this information, it will assign a score to your application. When recruiters look at the applications they have received, they will more than likely start with those that scored highest in the ATS. So, knowing how to elevate that score is crucial.

Section 6.2 covers how to choose the right job opportunities to apply to. But for now, we will assume you have found that perfect role—the one that you are an ideal fit for and really want to land. You know that you have the skills and experience necessary for the job, but now you need to make sure that the ATS is able to tell both the recruiter and the hiring manager the story you want told. That's why you are going to tailor your resume for the job posting, and you are going to do that by methodically walking through the requirements listed on the job posting and matching your resume to those requirements.

When you read through a job posting and specifically the list of requirements, you should be able to pretty easily pull out key words. These are what you need to focus on. The key words might be the name of a specific technology or skill. They might be a specific certification, degree, or other important facet. When you're tailoring your resume, you want to be looking for those key words and making sure that a match for them is found somewhere in your resume. Therefore, let's talk about a simple process you can use to make sure you maximize on these key words.

INVENTORY THE KEY TERMS

Your first step needs to be building a simple inventory of the key terms included in the job listing. Look at each requirement in the list; these are usually presented as a simple bulleted list, so this should not be too hard to locate. For each item, what is the primary thing the company is looking for, the main idea of that requirement, if you will?

Many times, it will be easy to look at a bulleted item in the list and pick out the key terms. Is it a technology like firewalls, containers, or cloud access security broker (CASB)? Maybe it is even a specific product name like Splunk, Kubernetes, or Burp Suite. Start making a list of these. For bullets that have a single key term (Burp Suite, for instance, being two words in the name of a single product), put that key term on its own line in your inventory.

In some cases, a single requirement will include multiple key terms. It might be a list of two or three similar products or a list of related technologies. When more than one key term occurs in the bullet, add each key term to your list as a separate line. While they may be included together on the job description for a reason, that does not mean you need to show them together in your resume. The goal is to make it easier for you to demonstrate that you have the key qualifications that the organization is looking for.

Table 6.1 shows a few examples of requirements taken from real-life job descriptions and the resulting key terms that would be added to the inventory.

Table 6.1 Example job requirements and keywords

Requirement from job description	Key term(s)
Strong working knowledge of vulnerability management practices and tools	Vulnerability management
Knowledge of provisioning, designing, constructing, and maintaining basic Azure compute instances	Azure

Table 6.1   Example job requirements and keywords *(continued)*

Requirement from job description	Key term(s)
Experience with regulatory compliance mandates such as ITAR, PCI, or HIPAA	ITAR PCI HIPAA
Experience with working with a SOAR management tool—for example, Demisto, Splunk, or Swimlane	SOAR Demisto Splunk Swimlane

As you work through this process, you may not always be sure of the key terms for a particular requirement. To help narrow it down, start by removing ancillary words that do not describe any particular skill, technology, or tangible quality.

Consider the first item in table 6.1. Straightaway we see the phrase *strong working knowledge of.* This phrase specifies the level of skill/experience needed but does not describe a skill itself. Therefore, you can eliminate those words. We also see *practices and tools* mentioned. Those words provide additional detail but do not, in and of themselves, describe a skill, technology, or quality. Once you eliminate those, you're left with the term *vulnerability management,* which is the key term in that requirement.

NOTE THE FREQUENCY OF KEY TERMS

When you complete this inventory, you will often find that some key terms are repeated more than once. This is a powerful indicator of the skills or experience the hiring manager values most in prospective candidates.

You can either list the duplicates while you are building your inventory and then match them up as a second step, or if you like to be more efficient, as you are building the inventory, add tick marks for each duplicate of a term that shows up. Either way, you want to identify those most crucial requirements if they exist. These are elements you will want to prioritize in the next step, as you start matching your resume to the key terms.

START MATCHING

Now it is time to compare your resume to your requirements inventory. Look at the knowledge, skills, and experience that you have included. Circle any keywords on the inventory that appear somewhere on your resume. These are cases where you have "checked the box." When the ATS scores your resume, it will recognize these keywords, and that will result in a higher score for your application.

Once you have circled all the key terms that match between your resume and that inventory, take a look at those that do not match. Are there key terms on the inventory that fit your knowledge, skills, or experience but perhaps are discussed differently on your resume? Reword wherever you can to ensure that the key words from the posting also appear in your resume.

Hopefully, this is obvious, but you need to be honest with yourself and your prospective employer as you do this. Do not simply fill your resume with key words describing capabilities you don't have. All you are trying to do here is optimize the wording of your resume to match up to what the employer is looking for. If you try to game the system by adding key terms that do not describe your capabilities, that will be easy for the recruiter to sniff out, and you can rest assured you won't be progressing in the hiring process.

As you reword your capabilities to match the key terms in the inventory, be sure to circle those that you have added. Now take a final look at the inventory and those key terms that you have not circled yet. Do any of them describe a capability from the capabilities inventory you made in chapter 4? Look for ways to include those in your resume. Again, any that you are able to honestly add, either as knowledge, skill, or experience, be sure to circle those on the requirements inventory.

ADD KEY TERM VARIATIONS

Finally, take a look at your resume and the key terms inventory. Are there key terms that you can list more than once? Perhaps you have experience with that skill at multiple previous roles. Look for ways to include that information, but also do so using words that are variations of the key term where possible. Most ATSs are smart systems. They understand and will rank you higher if a key term shows up in multiple ways. It is crucial to have the direct match first, but if you have related terminology as well, that shows greater overall capability, and the ATS will note that.

So, go through that inventory one more time. Maximize the mention of any of the key terms that you can on your resume. You'll end up with a resume that is tailored to the job description and makes it easy for the ATS and the recruiter to quickly identify that you have the requirements they are looking for.

6.1.4   *Proofreading*

The last step before you submit your tailored resume needs to be proofreading. Grammar and spelling errors not only look unprofessional to anyone reviewing your resume, but also can confuse the ATS and cause a lower score.

The ideal approach is to ask someone you trust to review your resume. This is a pretty standard writing tactic to ensure that your internal voice does not gloss over and fail to pick up on mistakes in your writing. This process can be more effective if you ask someone to review your resume who does not have skills in security or even IT. They will be more likely to pick out words and phrases that might be confusing to a recruiter who may not have security expertise either.

If you cannot have someone else review your resume, take a break after you are done revising it. Give it a few hours to sit (or overnight, if at all possible) and then come back to read it word-for-word from start to finish. Reading it out loud might feel awkward but can be particularly useful in this situation. Since you probably were not speaking your thoughts as you were typing, doing so now will cause you to reprocess your words and make it less likely that you will miss an obvious error.

6.2 *Choosing and applying to job openings*

Applying to jobs seems like it should be the easiest part of a job search in a market that claims to have millions of openings that will go unfilled. However, failing to find and apply to appropriate job postings is a common problem among entry-level applicants. Sometimes it is applying to jobs that they are not properly qualified for, but many times it's failing to understand when a job posting is worth applying to.

But the struggles do not stop there. Once deciding to apply for a role, applicants will often make crucial mistakes that result in their applications never making it past the ATS. Even if they do make it that far, believe it or not, applicants still make common mistakes that result in a failure to get an interview. So let's examine some of the strategies for addressing these issues.

6.2.1 *Using job search tools*

What do you do if you want to find a job in security? Most people typically begin looking through various job websites. They are convenient, enabling you to find many open positions aggregated in a single searchable database. Maybe you try two or three job sites. If you do, you will notice a lot of the same jobs posted in multiple places. You will also notice that some jobs posted on one are not on another. But you will not discover, of course, that some jobs will not be listed on any of them.

So let's take a step back and consider the moment before you even start searching for job listings on any site. You are about to take on a major challenge, finding a job in security as an entry-level candidate. Rushing in without a plan does not work well when you are getting ready to tackle any major hurdle in your career path. You have seen so far in this book that a methodical process can always help you organize and be more effective in your efforts. Job searches are no different. The way you go about searching for a job will be the foundation for a successful job hunt.

I mentioned previously job search websites. These can be powerful tools in your toolbox for searching for an open role that fits your skills. These sites aggregate open job listings from various organizations in a single, searchable database. Some are specialized to specific industries or even specific types of jobs within that industry. You can find job sites that focus on IT jobs, which include security roles. You can also find job sites that focus on security roles specifically. Even very specialized job sites exist that help you find positions in government jobs exclusively. Each site has its own database of jobs and its own way of matching candidates to job opportunities. An effective strategy, therefore, is to plan to use a variety of sources in your job search.

While these sites are great and bring together long lists of job openings from multiple companies, not all organizations use such job listing sites. They may not use them at all, may use only a small subset of them, or may list only certain jobs. So how do you find other openings?

Well, almost every organization has some form of career or job page on their website. While it is more time-consuming, it is helpful to your job search to put together a list of companies you know that are local to you or that you would like to work for whose sites you can go to specifically to see if they have open positions available. Some organizations even provide you the ability to apply to their company without applying to a specific job. While this is not a highly successful strategy, it is a good way to at least get your name and resume in their database and potentially get matched with future openings.

Other resources are available to you beyond just job search and corporate websites. If you are a recent or soon-to-be graduate, your school probably has resources to assist you in finding a job. Many universities have job boards and career placement services, and some even have relationships with local companies to help you find an internship to get you started in the field. These resources can be helpful in finding entry-level roles in particular. Since their focus is on recent graduates, they often have better access or awareness of where entry-level roles can be found.

Of course, if you're looking to pivot from another career path and are not a recent or soon-to-be graduate, you may not have access to these sources. Still, it cannot hurt to at least check with local universities to see whether they have any publicly available services like this. Some schools do indeed offer at least some level of assistance.

Within the industry itself, recruiters can work on your behalf to help you find a job. These recruiters work on fees that are paid by prospective employers. These types of recruiter services can be helpful by exposing you to many open roles, and they do much of the work in assisting you with finding roles you are qualified for.

Unfortunately, the flip side is that working with these recruiters can introduce an additional layer of scrutiny on your skill set that you may have to work to win over. Recruiters who work with these services typically make a commission when a candidate is successfully placed. Therefore, while they do work as your advocate and want to place you in a job, they sometimes are less likely to take a chance on roles. They often will look to submit you to job openings that have a high likelihood of successful placement. That is great unless you are an entry-level person looking for scarce entry-level jobs. Sometimes it can be more effective to take a chance on a role and try to earn the opportunity to make your case.

Finally, let's not forget about social media. Hashtags, such as #infosecjobs, often can be helpful in locating posts or tweets about available jobs. Within the security community itself, you can often find advocates who will start threads asking their followers to post open job opportunities. If you have started to build a social network, you can even reach out to some of these advocates who might gladly share your information and your needs for a job with their followers to help amplify your reach. While I wouldn't consider these methods to be primary in your job search, you should at least be plugged in and paying attention.

Taking this all into account, it will help your search to have a plan for which methods of searching you want to start with. Ultimately, your success will depend in part on how much you leverage the various tools at your disposal. Have a plan for which resources you will start with. Make sure it is a strategy that uses a diverse set of these resources to maximize the chances of your success. Now that you know where you are going to look, let's talk more about how you can find the right jobs.

6.2.2   *Finding the right roles*

In chapter 4, you did a lot of work to identify your interests and the types of security roles you might enjoy. That information is a great place to begin. If you are going to go searching for jobs, you definitely want to know what you are looking for and how to easily find it. The simple fact is, companies don't use standardized titles for their security jobs (or any other for that matter). For instance, you may see a penetration tester role listed as a penetration tester, security analyst, ethical hacker, or security tester. Many of the big job sites will attempt, to varying degrees, to help match your keyword searches to related key words like that, but they tend to be inconsistent.

Trying to use search tools on job websites can be frustrating as a result. The key is to understand and accept that there is no perfect search and that you will need to conduct multiple searches to reveal as many of the right roles as possible. Searching for multiple possible titles is one way to get a comprehensive list. However, make sure your searches extend beyond just job titles. Look at your capabilities inventory from chapter 4 and set up searches that look for your key skills or experience. This will also help you build a longer list of potential strong fits and can actually be more effective than searching for a specific job title.

Location, location, location

Location searches on job sites can be particularly erratic, so it is important to understand a few concepts here. Many sites force employers to list a location in their job opening listings. Some allow only one location, while others allow multiple. Some sites allow job postings to be listed as remote (to work from home), while other sites do not. Some sites allow job seekers to search for remote jobs, and others do not. Some force you to include a search location, while others do not.

Remote roles have become common in the security industry. While they are less common for entry-level positions, they do still exist, and you may be interested in searching for these kinds of roles.

To get past the limitations of some of these job sites, companies may list the same job in multiple locations. If you see the same job posting for a particular company repeated a bunch of times with different locations, this could be an indicator that they are open to hiring the role in any of multiple offices they have, or it can even be a sign that the position is open for remote work. When you are searching, keep this in mind as well.

Vary, as much as the site will allow, the locations you use to search. Perhaps use just your state. If you live close to a state line, maybe include the nearby state. If you can search without a location, try that too, but remember to be more specific than in your other search terms to ensure that you don't end up with an unwieldy list of potential openings.

As with just about every topic we will discuss, I suggest you use a methodical approach. Create a list of job titles you will search for. Also, go through your capabilities list and pick the ones that you feel are strongest or most suited to the kind of role you would like to land. Start with any experience-level capabilities you have, and then fall back to skills, and if necessary, pull in some from your knowledge list. This gives you a great way to plan your searches and make sure your searches return roles that will be a good fit for your desires and capabilities.

6.2.3   *Matching requirements*

As you do your searches, start reading through the lists of results and picking the opportunities that sound interesting. Maybe you are drawn to some that are in your area or with a company you have heard of, or maybe just the title suggests it would be a good entry-level opportunity. Now you start opening them up and you are bombarded with a list of requirements, and your task becomes more daunting. You may get frustrated as it begins to feel like you are not qualified for any of the jobs you are finding.

A common response to feeling overwhelmed in a situation like this is to shut down and not bother applying to any. You might try applying to a few that seem closest to your skills but don't really feel like a fit. Or maybe you are the type to just apply to everything and hope for the best, knowing the worst they can tell you is no. However, all of this seems rather random and lacking in any form of methodology or structured approach. There has got to be a better way, right? Well of course there is; let's take a look.

Remember that capabilities inventory you created in chapter 4? Well, that is about to come in handy again. Not just your completed inventory, but the methods that you used to create it. Job descriptions can be complex, and that is when they cause problems for candidates applying to them. Therefore, when evaluating a job listing to determine whether you are qualified, you need to be able to break down the list of requirements and determine which requirements are actually the most crucial for the role.

You have conducted your search and found a few job listings that you feel would be interesting to you and at a high level match up to your capabilities. Now it is important to understand the structure of a job description. Typically, job descriptions have a few common sections. Of course, you'll see the basics like title and location. Often you'll also see a high-level description of the role. A lot of times this is accompanied by a responsibilities section, which is usually a bulleted list of the day-to-day duties that the role is expected to fulfill.

However, it is the next two sections that we want to focus on here. They are the requirements and the additional qualifications sections. The requirements section should list the minimum qualifications that a candidate must meet to be considered for the role. This section might be labeled Requirements, Minimum Qualifications, or another similar title. The additional qualifications section lists other capabilities that would be helpful in making a candidate successful in the position. These are items that the candidate is not expected to have, but candidates who do demonstrate them will be considered more qualified. This section is also sometimes labeled as Preferred Experience, Additional Skills, or something similar.

Those are the ideal scenarios, but job descriptions often do not meet those ideals. You may come across job descriptions with a long and intimidating requirements section. This is usually a sign of a company that is including more than just minimum requirements in that section. You may also see requirements that list a specific qualification but then include the notorious phrase "or equivalent." For instance, you may notice things like "Bachelor's degree or equivalent work experience" or "CISSP or equivalent security certification." The goal when considering these requirements is to figure out which of them are the most important, and you can use a few clues to make that determination.

First, consider the order. While it is not always an indicator, think about how a hiring manager or human resources team might put this job description together. Generally, they are going to sit down, think about the role, and attempt to identify a list of requirements. The first ones they think of will come to mind for one of two reasons. Either they are easy general elements, like degrees and certifications, or they are the first skills that come to mind when someone thinks of that specific role. So, if you encounter a long list of requirements, give priority to the first ones in the list.

Another clue in determining which requirements are most crucial is to examine how they link back to the description and responsibilities listed in the job description. Look for themes in the terms being used. If necessary, write down each requirement and then note how many times something related to it appears in the description or responsibilities section. The more specific the requirement is, the more narrowly worded it is, the less likely it is to be significant in the decision-making process.

Finally, look for relationships between the requirements list and the additional or preferred qualifications list. Many times, you will notice a high-level concept in the requirements that is backed up by the mention of specific technologies or skill sets in the additional qualifications section. For instance, you might see a requirement that says, "Knowledge of cloud security best practices." Then in the additional qualifications section, you might see something to the effect of "Experience with Google Cloud or Microsoft Azure cloud architecture." That is a good indicator that capability with cloud technology is particularly important for this role.

One last word about job requirements. While you want to check off the boxes for as many of the requirements as possible, avoid feeling like you need to check off every single item in order to be qualified to apply. Remember, first and foremost, the worst they can tell you is no. I am not suggesting you should apply to every job you see; you do still need to be tactical about which jobs you apply to. However, it is unfortunately common for job applicants to self-eliminate themselves from jobs they were qualified for simply because they failed to meet every single requirement on a list.

The more requirements there are in a job description, the less likely it is that a candidate will come in that matches all of them. So do not fall into the trap of assuming that someone out there will match all of them and therefore feel like you should not even try.

6.2.4 *Office, remote, or hybrid working environments*

Many people were introduced to working from home through the coronavirus pandemic. As companies were forced to close offices, many workers suddenly had to find space in their homes to set up a mobile office and continue to perform their jobs over remote connectivity solutions like virtual private networks (VPNs) or virtual workspaces.

As offices started to reopen, many employees found they preferred the remote working experience. In response, organizations converted many roles to fully remote, while others opted for hybrid models that leverage different arrangements of working part of the time in the office and part of the time at home.

These shifts in working environments have now become an important consideration in the job search process. For some, the idea of working fully remote may sound wonderful. No long commute into the office, comfortable and familiar surroundings with fewer distractions, and even the added flexibility in scheduling personal appointments can be attractive options. However, for others, the idea of being at home all the time is untenable. Many who thought they would enjoy a work-from-home environment have found that they miss aspects of being in the office, such as interacting face-to-face with coworkers or having easy access to other facilities or equipment.

Therefore, it is important to take an honest look at yourself and examine what environment might work best for you. Are you disciplined enough to avoid distractions that can occur in a home-based office? Do you have a space you can dedicate to your workday? Do you have the necessary infrastructure to work from home, including a sufficiently high-capacity and reliable internet connection and phone service? These are important considerations for a fully remote or hybrid working environment.

Considerations for a home working space

The following is a list of items and best practices that you need to consider if you are going to pursue a remote or hybrid working environment:

- A dedicated space for conducting work. It could be a full room, maybe just a desk in an out-of-the-way corner of your space. Whatever it is, look to make it a permanent dedicated space if at all possible (not the kitchen table).
- Thinking still about space, depending on the role, you may need a secluded space where others won't be able to overhear you discussing sensitive information. Take this into account as well.
- Stable and fast internet service is a must. Many of the systems you will connect to will need significant amounts of data transfer, much more than just browsing a website. Video conferencing and other tools in particular need stable and fast connections.
- Stable phone service will also be crucial in some roles. Do you have or can you add a landline phone? Do you have dependable cellular service? Some employers are able to provide software phones that work via your computer.

> - Will you need the ability to print, and if so, do you have a printer? This often goes overlooked, but many people do not have printers anymore. If you are not in the office but are expected to be able to print certain forms or documents, this could be a consideration as well.

On the flip side, however, working fully remote can open up more possibilities in your job search. If the company you are applying to is located three states over, obviously they would not be an option without either relocating or setting up a remote working arrangement. Additionally, some people who are disciplined and independent workers thrive in a home working environment. In the end, you need to consider the best option for you.

6.3 Crushing the job interviews

You made it—well, at least to this point. You have learned about the roles in cybersecurity. You have picked an area or two of security that you want to focus on. You have completed your personal capabilities inventory and self-analysis. You have applied to a few jobs. And now you got that email from one of the recruiters saying they are interested and would like to schedule an initial interview!

All the work you have put in so far has led to this point: your opportunity to show them why you are the person they want to hire, the person that they must hire, for this job. However, it is also important to remember that you are interviewing the organization as well. This is your opportunity to figure out whether the job, the company culture, and the compensation offered are a good fit for you. Unfortunately, it can be easy to become so focused on trying to impress that you forget that they need to impress you too.

At this point, you are probably feeling a mix of excitement and anxiety. This is a high-pressure process; you want to say and do all the right things to make sure that you win the job. You are getting ready to learn about a company you could be spending the next 2, 4, 10, or more years at. It is a big moment in your life, no doubt. So, let's spend some time talking about the interview process.

Every organization has its own approach to interviewing potential candidates. Some may have just a few informal conversations, while others might require a marathon of interviews and even possibly technical skills evaluations. However, while every one of them is different, you can prepare for common practices in the interview process.

The overwhelming majority of organizations start off the process with the human resources screen. This is then usually followed by an interview with the hiring manager. From there, you may run into a mix of technical interviews, group interviews, and possibly even requirements for you to demonstrate practical application of key job skills. You might meet with team members who would be your peers on the team, potential peers from other parts of the organization, and even possibly different levels of management. To be successful, understanding the format, goals, and ways to prepare for each is crucial.

6.3.1 *The recruiter or HR screen*

Almost invariably, the first conversation you will have once your application is accepted is with the recruiter. They serve as a first line of screening for potential applicants. At the point the recruiter is reaching out to you for an initial screen, they, and likely the hiring manager as well, have reviewed your resume and decided to explore your fit for the role. The job of the recruiter at this point is to look for any administrative issues or deficiencies in your skills and experience that would disqualify you for the position.

The recruiter screen is typically a fairly short conversation, usually 30 minutes or less. These conversations are almost always conducted by phone. You can expect the interviewer to spend some time telling you about the organization and the position you applied for, and offering to answer any initial questions you have.

The recruiter will likely ask you some administrative questions about your availability to work, your job history, and possibly salary requirements. Depending on the organization, they may ask you more detailed information about your skills, but usually that is left to the hiring manager and other subsequent interviews. They should share with you information about benefits and other forms of compensation that would be a part of the job.

Finally, they should share with you a summary of the interview process from start to finish. They may or may not tell you whether they plan to move you on to the next step of the process; however, it is pretty common for interviewers to avoid making any commitment about next steps. Usually an internal review involving other people in the organization occurs before the recruiter commits to next steps.

Based on this, you should come to the recruiter screen ready to discuss your employment history, your skills, and reasons you are applying to this job. It is not uncommon to be asked why you are looking to change jobs, so have a good, honest answer prepared for that question. This should be a relaxed interview with most organizations. The recruiter is just trying to figure out whether you have enough of the required skills to meet the minimum needs of the job. They are likely comparing you to other applicants and will pass along those who are qualified, for the hiring manager to make a decision as far as subsequent interviews.

You should also come prepared with your questions about the organization, high-level questions about benefits, and even questions about the role itself. The interviewer may defer some of those questions to the hiring manager, but it is always better to ask and be redirected than to hold back and not get important information. Also, if the recruiter does not share information about the interview process or what to expect, make sure you ask. That information should not be secret.

As I said earlier, this is also your opportunity to interview the organization. Just as they are trying to determine your fit for their job, you should be trying to determine whether the organization is a fit for your needs. Pay attention to what the recruiter tells you about the company. Why are they hiring for this position? Is it a new role that came about because they are expanding a team? Did someone in the role move up into a higher-level role? These are indicators that can tell you a lot about the organization. Also, notice how structured or casual the conversation feels—that can give you a sense of the corporate culture within the company.

To prepare for the interview, make sure you have researched the organization and understand their business. Know what products or services they offer. If you can, learn about the team that you will be joining and how it fits into the overall business. It is a good idea to look at websites that provide employer reviews as well. See what comments current and former employees are posting about the company. This might lead to questions you want to ask in the interview.

If you have the time, it can even be a good idea to search social media (for example, LinkedIn) for other people who work at the organization to identify key players in the leadership team. Showing familiarity with the organization from the initial screen and throughout the process can make you stand out as a candidate. It shows that you are not only passionate about the job, but also organized and prepared.

Make sure at the end of the interview that you know what to expect next. Chances are they will tell you that they are going to discuss you internally and will contact you regarding the next steps. They should provide you with an approximate time frame for when you should hear back. If they do not, certainly make sure you ask. You should expect to get an answer either way, and you should know when the right time to follow up would be if you have not received an answer.

The salary question

It is common, especially in the initial recruiter screen and possibly in subsequent interviews, to be asked about your salary requirements for the job. What they are asking is the amount you expect to make in this job.

What they should not ask you is the amount you are making now. While this used to be a common question, it is a practice that is losing favor among recruiters and, in many locations, it is illegal for prospective employers to ask about salary history at all. You should become familiar with the local laws governing this, and if you are asked about your salary history, politely redirect the conversation to talking about what you expect to make in the role you have applied to.

You should go into the interview with a clear idea of the salary you expect to receive. Unfortunately, while organizations may be pressured to be open about their salary ranges for jobs, few freely offer this information. You can do some research on your own. Job sites sometimes post salary ranges based on jobs of similar types. Sometimes these are based on general salary surveys, but some sites draw the information from people who have worked for that specific organization.

Use that salary range information in conjunction with an objective look at your own qualifications for the job to pick a realistic number to ask for. Remember, this number

(continued)

will turn into a negotiating point if you do get offered the job, so the trick is to ask for an amount that is realistic for them based both on the salary range and your qualifications, but also maximizes your potential earning if you do land the role.

6.3.2   *The hiring manager interview*

In many organizations, the next interview after the recruiter screen is with the hiring manager. This is your opportunity to really shine; after all, this is the person who will ultimately make the decision of whether to hire you. It is also your chance to get to know and evaluate the person who will be your manager if you are hired. While you want to be focused on presenting a confident and professional demeanor, you also want to interact fairly casually to get a feel for how day-to-day interactions may be handled.

The hiring manager will almost assuredly dig deeper into your job history, your skills as they relate to the job, and your overall approach to the job. They will likely tell you more details about expectations for the role, their approach to managing the team, and the skill sets that are most important in the job.

You should expect a healthy dose of questions about your skills and work experience, but you should also have time to ask your questions. In this interview, you should ask specific questions about responsibilities within the role that were unclear from the job description. You may also want to ask about the inner relations of peers within the team and how the team fits into the overall organization.

Preparing for this interview is important. As soon as you have been scheduled for an interview with the hiring manager, look them up on social media. Use LinkedIn, in particular, to find out more about them, their background, and their views on security-related topics if they post about them. Look for commonalities you have or areas you can ask more about. If they have videos of public speaking engagements, articles, or blog posts, check those out as well to get to know your (hopefully) future manager.

This information can help you understand their potential motivations behind certain questions they ask. It can also help you better craft your answers, to speak more specifically to their beliefs. A word of caution, however: I am not suggesting you placate or answer questions dishonestly to try to match their beliefs. Be careful about referencing their materials in your discussion. Once or twice on appropriate topics is good, but if you try to force it or do it too much, you may come off more like a stalker than an organized and informed candidate.

By the end of this interview, you should have a good feel for what the job will entail on a daily basis and how you and your potential boss will mesh from an interpersonal perspective. Before ending the interview, make sure that, again, you get information on next steps, when you should expect to hear back from them, and who the appropriate person is to follow up with (more than likely it will be the recruiter, not the hiring manager).

Strategic questioning

Come to every job interview prepared to ask questions. Asking the right questions can sometimes be even more effective than showing off your skills and experience. Asking questions that demonstrate you understand their business and/or the role you are interviewing for can make you stand out.

Asking good questions shows that you are already thinking proactively about the job and potential issues or challenges that may surface after you have begun working for them.

For instance, if you have experience with specific technologies included in the job description, you could ask about specific features the organization is using or if they are experiencing any challenges with the tools. If the job requirements mention any specific security frameworks or regulatory concerns, you could dig further into how well they have been able to meet those requirements.

Again, this is a great way to stand out among candidates and exhibit both your understanding of the job as well as your passion and excitement. In my experience as a hiring manager, I have always paid particular attention to the questions my candidates ask. It gives me visibility into how dedicated they are, how organized they are, and how they think about given situations.

6.3.3 *The technical interview*

After making it past the recruiter screen and the hiring manager interview, you will likely be asked to go through a technical interview. Sometimes these can even occur before the hiring manager interview. Technical interviews cause some of the highest levels of anxiety among job candidates. You know that your skills will be put to the test and you will more than likely be talking to someone or multiple people who have a higher degree of experience and skill than you. That can be daunting, but do not let it scare you off. Remember, you got this far because of the things you have done and the skills you have developed that you shared on your resume.

In a technical interview, as I mentioned, you may meet with one or more members of the organization. They might be members of the team that you are looking to join, but interviewers also could come from other areas of the organization. Technical interviews, as the name indicates, are meant to measure your technical aptitude for the job. If it is a group interview, they may also be looking to assess your fit with the team. The key, no matter what, is to stay relaxed, keep it as casual as you can, and try to enjoy the experience.

The nature of the interview will vary based on the organization. Some have a structured feel. The interviewers might be working from a list of questions and taking notes, potentially even scoring you on your responses. In other cases, the process may be more casual, and the questions may be more about hypothetical situations than direct quiz-style questions. The way you respond will depend on the type of questions they ask.

If you find yourself in a technical interview that includes direct test-type questions, craft your answers to be succinct and direct. Be sure to provide enough detail to effectively answer the questions, but do not go off on a tangent or try to fill time on questions that you feel you are not fully equipped to answer. Provide as much of an answer as you can while being honest about your limitations.

If you instead find yourself in a technical interview with questions that are more hypothetical, sometimes beginning with "Tell me about a time" or "How would you handle this situation," you'll want to offer a more detailed answer. From these *behavioral-based questions*, interviewers are looking for you to demonstrate your skill while also sharing more about your personal behaviors. One method for answering these types of questions is the STAR method.

Figure 6.1 The STAR method for answering behavioral-based questions

The STAR method provides a simple framework for constructing your answer to these types of questions. It is most effective when you are drawing from a past experience, but you can also use it when answering a hypothetical situation. The goal is to draw a mental picture for the interviewer of how you approached a past situation or would approach a situation if it occurs. The process includes the following:

- *Situation*—Quickly provide context for the situation you are going to use.
- *Task*—Describe the problem or expectations.
- *Action*—Provide details on what you did or would do.
- *Results*—Share the outcomes or the expected outcomes.

Preparing for a technical interview can be daunting. To help make it easier, when you get scheduled for the technical interview, ask the recruiter if they can share suggestions for how to prepare or the kinds of questions you should expect. I did this once while interviewing for a position and the recruiter ended up scheduling a 30-minute prep call with me that armed me with a lot of information on what to expect and how I could prepare. Asking for this type of information again shows that you care about the job, that you are approaching the process in an organized and detail-oriented fashion, and of course it can also help you better prepare for the interview.

You should also spend some time researching the people you will be interviewing with. Again, use social media and other searches to find out their backgrounds, the topics or technologies they have interest in, and the way they approach various technical problems. Knowing the background of your interviewers can help you prepare for the kinds of questions they're likely going to ask as well as help you avoid delving into a topic that's over your head.

Technical interviews can be particularly exhausting. Sometimes they can last more than an hour. Approach these interviews as you would a big final exam. Get plenty of sleep, set up your day to avoid outside stress factors, and do your best to come in relaxed and focused.

Even though the interview is focused on measuring your skills in an objective fashion, you will likely have the opportunity to again ask questions. This is a good time to get more information about your potential manager if you are speaking to peers on the team. Ask questions about a day-in-the-life of someone in the role you're interviewing for. Ask direct questions about how the team works together, how the manager interacts with the team, and even the challenges or frustrations they experience. This is one of the best times to get a clear view of the culture within the organization and the team.

Before leaving the interview, you should again be given details on the next steps and when you may expect to hear back. If the interviewers don't have this information, you may need to follow up with the recruiter. Either way, be sure you know when to expect an answer so you know when it would be appropriate to follow up.

When you just don't know

Honesty is not just an ethical imperative for the interview process; it can be an effective tool in making you stand out. In a technical interview, for instance, when you don't know the answer to a question, being honest that you do not have a good answer is preferable to trying to fake your way through it. Good interviewers can pick up on that type of behavior pretty quickly and will know that you are trying to talk around the subject. So instead, if you are unsure of the answer, be up front about it. However, then you can take that opportunity to expand and try to use the knowledge you do have to figure out what the answer is logically.

For example, in one technical interview I was asked about a certain type of software attack called Prototype Pollution. I had never heard of it before. To answer the question, I started off by admitting right away that I did not know for sure what that was. I then explained that I know that in software development, a prototype is the definition of an object or a method. I continued to explain that the word pollution suggests that someone is maliciously modifying the prototype or injecting malicious content into the prototype.

This proved to be an effective answer, as it showed that while I didn't know the exact details of that type of attack, I had enough background knowledge to learn about it. Also, I was able to demonstrate the capability to think about a problem logically and come to a reasonable resolution. After I ultimately was hired for the job, the person who asked that question told me how impressed he had been with the answer.

6.4 Considering the job offer

It happened. After all that work, surviving round after round of interviews, you were successful, and you got the phone call telling you they want to hire you. Perhaps they have already sent you an email with a formal job offer. Your work is not over yet, as you may still need to negotiate to make sure that you get the best possible deal you deserve.

6.4.1   *Don't rush things*

First things first—when you get that phone call telling you that they would like to move ahead with hiring you, there is no reason you have to say yes. If you came this far and have reservations about the job, the people, the culture, or whatever, there is no reason you should feel that you are now committed to accepting the position. When the call comes, it may be the recruiter, the hiring manager, or both who will contact you. They will likely describe the offer in terms of salary, any form of bonus program, and the benefits like vacation and sick days. Most companies will send you a formal offer letter with all the details as well.

When you do get that call, be gracious and thank them for the offer and the opportunity. However, I strongly suggest you ask for time to consider their offer. You should not feel like asking for this time makes you seem ungrateful or uninterested. You can even tell them that you are excited but just want a day or two to read through the offer (if they will be supplying a written offer) or to consider the details and get your thoughts in order before you officially accept.

If you have reservations about the job or the offer they have made, this gives you time to organize your thoughts and decide whether you want to try to negotiate a better deal or to outright decline the offer. If you are totally happy with the offer, it is still good practice to take time to relax and fully consider it. Your emotions will be likely running high, and taking a step back can help make sure you do not run full-speed ahead into a situation that was not what you were expecting.

If they send you an offer letter, take the time to read it through in its entirety. Understand the salary, other compensation (including any bonus package or signing bonus), benefits, and other matters. If you have specific questions about their benefits package, this is the time to get those answered. Take at least a few hours or even a day or two to consider the offer and make sure it is the one you want. Set the expectation for how much time you need, but remember that they may have other candidates for the job that they haven't provided an answer to yet because they are waiting for you to accept or decline. Generally, one to two business days is acceptable for considering an offer before providing a response.

6.4.2   *Negotiating for something better*

A job offer is not a final offer. Kirsten Renner, a recruiter whose focus is on hiring for cybersecurity roles, puts it this way: "Everything is negotiable." If you are not happy with the salary offered, you can ask for more. If the number of vacation days offered is too few, ask for a better package. Ultimately, you want to make sure that you are comfortable with the offer, not settling for something that does not meet your needs.

When negotiating, keep in mind, however, that it is a give-and-take process. Sometimes the company cannot change certain factors but can offer you a concession in other ways. For instance, perhaps the salary offered is closer to the top of their salary range for that job and they are reluctant to go any higher. In this case, perhaps a signing bonus or additional vacation days or a higher bonus payout percentage can be set up instead to make up the difference.

If you decide you want to negotiate for a better deal, make sure you know what is most important to you going in. Be realistic and understand you may not get everything you are looking for. However, on the flip side, do not be afraid to at least ask. The worst thing you can do is to accept an offer you are not happy with because you are afraid they will rescind the offer and hire someone else. That situation rarely happens, and when it does, it is usually the sign of a toxic culture you probably would not want to work in anyway.

The negotiating process needs to begin with you making it clear which parts of the offer you are unhappy with and providing a reasonable proposal. Do not attempt to start negotiating your offer by rejecting the offer. Once you have declined their offer, they have no commitment to you, and most companies will walk away at that point. You might get lucky, and they will inquire as to why you rejected it, but that is a big risk to take. Instead, your response should be gracious, thanking them for their offer and then explaining your reservations and how you would like to see that remedied.

The key to a successful pre-employment negotiation is being reasonable and cooperative. Be confident in your worth and make sure you get paid for it, but avoid coming off as arrogant or unrealistic. That will quickly shut down the negotiations. Remember, if you are successful, these are people you will be working with every day. So, be firm but polite and professional.

6.4.3   *Employment agreements*

One last thing to be aware of in your offer, or to ask about if it is not mentioned, is whether the company requires any form of employment agreement. Many organizations require their employees to sign an employment agreement that might contain clauses for noncompete and nonsolicitation.

Noncompete clauses typically state that for a period of time after you leave the organization, you are not allowed to work for any competitors of the organization. *Nonsolicitation* clauses typically state that for a period of time after you leave the organization, you are not allowed to attempt to influence their employees to leave the organization. Other clauses may constrain the way you may or may not interact with the company's customers for a set period of time as well. Be aware of these restrictions, as they can be limiting if, and when, you decide to leave the company down the road.

Terms of an employment agreement are typically not negotiable, as everyone within the company is usually required to sign them. You can try, if you are not comfortable with a specific element, but employers rarely make exceptions. Also be aware of local laws. Some locations do not allow noncompete clauses, so that portion of the agreement may not be enforceable. It is ultimately a contract you are signing; if you are concerned, it is prudent to have legal counsel review the agreement and provide guidance before you sign.

Summary

- When building your resume, you should ensure that it is formatted in a way that an ATS can easily parse it and includes keywords that can easily be matched by the automated system to the requirements of the job description.
- Choosing the right jobs to apply to requires you to evaluate your personal objectives and match your skills to the job requirements. Additionally, analyze job requirements and identify the ones that the hiring manager sees as most important.
- Core skills should be represented on your resume in a way that truthfully links them to the job requirements. Show how skills from a nonsecurity job fit the requirements for specific skills in the job description.
- Candidates should prepare for each interview based on the type of interview being conducted. Proactive research and preparation are crucial.
- Avoid pitfalls such as self-elimination from potential jobs, being dishonest, or attempting to cover up gaps in technical knowledge.

Part 3

Building for
long-term success

At this point, you might be wondering, what else is there to discuss? You have learned about the cybersecurity field and options for various career paths. You have self-analyzed to find the best fit and have armed yourself with the knowledge necessary to be successful in your job search. However, the last thing anyone would want to do is launch a new career path and then have it not last. If you are going to put all this effort into becoming a cybersecurity professional, you probably have a strong interest in making sure it is a lasting journey. That is what we are going to cover in this third and final part of this book.

Chapter 7 kicks off by highlighting the need for mentorship and networking. You will find topics that arm you with strategies for building a professional network, leveraging various tools at your disposal. You will also discover proven techniques for finding a mentor and understanding which types of mentorship will be most valuable to you.

Chapter 8 continues the focus on long-term success with a full analysis of the single most career-limiting challenge you will most certainly encounter: imposter syndrome. This chapter not only defines imposter syndrome and its causes, but also offers practical tools to help you manage its impact and overcome the negative force that it threatens to have on your career growth.

Finally, as we finish off this guide, achieving long-term success becomes the goal. Chapter 9 covers practical goal setting for your career journey. You will also learn about pivoting among disciplines in cybersecurity and become prepared to make those jumps with confidence. The guide concludes by helping you take all the knowledge you have gained and put it into action.

You are almost there. Of course, when you finish reading this guide, you are not finished. I truly believe this guide will be helpful to you for many years to come. Now settle in and get ready to learn the keys to maximizing the success and longevity of your cybersecurity career. This is what you have been working toward. The time has come.

The power of networking and mentorship 7

This chapter covers

- Leveraging tools to build a professional network of security professionals
- Finding a mentor by identifying the right characteristics and traits
- Setting up a mentoring relationship based on the right expectations
- Gracefully ending a mentoring relationship

In chapter 6, we spent a great deal of time talking about preparing for your job hunt, maximizing your chances of getting a job interview, and negotiating effectively once you get the job offer. These strategies should help you find and land a rewarding position, but sometimes you could use additional help. This is where the power of your network and professional development via mentorship come into play.

To better illustrate the importance of having a strong network, let me share a little from my own journey. I recently moved into a new role that saw me achieve some of my highest goals as a security professional, but I didn't get the job simply by searching job postings online and applying. The role wasn't even posted online at all. Instead, I found out about it through a CISO whom I had interacted with on multiple occasions through LinkedIn, Twitter, and even a virtual roundtable event. It was in talking with her about her role, and also my experience and goals, that the opportunity came to light. She connected me with the hiring manager, and as soon as the job was posted officially, I was entered into the interview process.

An old adage states that it's not *what* you know, but *who* you know, that matters. When it comes to finding jobs in cybersecurity, this can often hold true. Having a powerful network of professionals whom you interact with through social media, meetups, or other events can lead to opportunities just like mine.

That strong network can do more than lead you to new job opportunities. It can also help connect you with other professionals who could serve as mentors to you. Mentors can be found in many places and don't even have to be in the same field. They help with professional development, giving you perspective and knowledge to truly excel in that cybersecurity role you are looking to land. So, let's look deeper at what networking and mentoring can do for you and how to effectively develop and leverage both.

7.1   Building a professional network

Spend some time on a professional social media platform like LinkedIn and in a short amount of time, you will likely see talk of networking events. Indeed, in the business world, professional networking is a practice that receives a lot of discussion and attention. Professionals meeting with other professionals, building relationships, and learning from each other is a powerful tool for anyone looking to grow their career.

However, for the first-time job seeker looking to land a role in a new industry, the task of building and cultivating a network can be daunting. It seems so much easier to connect with people in a particular industry when you're already working in that industry. When you feel like you are on the outside looking in, it is difficult to even know where to start. Fortunately, you can employ certain practices and techniques to overcome that hurdle.

7.1.1   Social media

Social media has become a powerful force in so many ways in our lives. When you start thinking about building a professional network, there is a good chance that leveraging these platforms comes right to mind. And that's good, because social media is an easy way to get started and bypasses many of the barriers that we will discuss later in this section. On the flip side, however, social media presents challenges that you need to understand and possibly overcome in order to use it effectively in building your network.

One of the great things about social media is how easily it can allow you to connect with people all over the world who have interests similar to yours. In this case, your interest is cybersecurity, or perhaps you even want to focus on a specific discipline within cybersecurity that you are most interested in pursuing (perhaps those you identified in your personal objective statement in chapter 4).

Whatever you decide, the key is to be intentional about building a network through social media. This is not something you want to leave to chance, or hope happens via some serendipitous event. If you need to, take a little time and really think about the cybersecurity interests that you want your network to be focused on, and plan for connecting with those people who have similar interests.

But how do you know who to follow, and how do you get connected with them? Regardless of the platform, one of the easiest ways to get started is to identify some relatively well-known people and follow or connect with them. It might be a cybersecurity reporter who authored an article on the topic or interest area you have chosen. It could be a popular personality in the security community who gave a talk or is known to specialize in that particular subject matter. Regardless, the goal is not so much to see what they have to say and share, but to pay attention to who is commenting on it and resharing those materials.

Paying attention to the responses and reshares/retweets provides you with two things. First, it identifies for you other people with a similar interest in the same topic whom you may want to start following as well. This is an easy way to start filling your feeds on the various platforms with commentary from people you may want to network with. Second, it gives you the opportunity to interact. A great way to start building a network is to simply respond and provide your own thoughts or ask questions about the original post. Even if the original author does not respond to you, someone else who follows them might. Now you have the opportunity to talk more and possibly get to know that person and ultimately build that connection.

The key with social media networking is ultimately all about interaction. The more you put your name out there in the cyberspaces, the more people will see you and recognize you as someone with interest in those topics. As you follow more people and their posts end up in your feed, the more you will get to know them as well.

You might even see someone post something about a job opening that they have. If they know you well enough and know you are looking for work, some might even seek you out. As you build those connections, you in turn can find other people with whom you might start to interact. It can happen relatively quickly if you're able to dedicate just a couple of hours a week to really interacting with the professionals in this field.

7.1.2   *Industry groups and networking events*

As I discussed earlier, the business world places a lot of focus on professional networking. Cybersecurity is no exception. Many groups and organizations are dedicated to providing networking and knowledge-sharing opportunities for people working in or interested in the industry.

Different groups have different purposes. Some are general professional organizations that focus on cybersecurity at a high level. Others have technical interests and focus on sharing and growing the skills of their members. Still others focus on specific needs of our community, such as bringing more women or people of color into the organization. The one thing that most of them share, however, is their openness to welcoming new members.

This is another area where having begun building your professional network on social media can help. Following and connecting with various members of the industry will no doubt result in you seeing content regarding meetups and other networking events that are occurring. This will help you identify some of the organizations that you might want to consider becoming involved with yourself. National and even international organizations may have local chapters that you can join. Depending on your location, there may even be more local organizations you can connect with. It is a good idea to explore the various available options and find the ones that have events or a focus that feel most interesting to you.

One of the challenges is that these types of groups can be difficult to navigate for people who are introverted, deal with social awkwardness, or even have more severe social anxieties. However, if that description fits you, it does not mean that these groups can't still be of value. Perhaps try looking for a local chapter or group that focuses more on skill-building events. They might host CTF competitions or have presenters who give a workshop on a particular tool or technique. For some, this type of event can provide an easy icebreaker to connect with a few people without the pressure of a full social situation.

While many local groups might be specific to your area, the following are a few international organizations focused on security that might have a local chapter nearby that you could get involved with.

DEF CON GROUPS

Spawned from the hacker conference DEF CON, *DEF CON Groups* are individually run local groups that can be found around the world. While each has its own approach to hosting meetings and to the activities involved, they are a great way to meet with fellow security enthusiasts, professionals, and hackers. Within the United States, the groups are referenced by their area code; for instance, DC414 is the local DEF CON group in Milwaukee, Wisconsin. Internationally, telephone codes are still used but obviously vary by country. Generally, attending a DEF CON Group meeting costs nothing.

OWASP

The *Open Web Application Security Project* (*OWASP*) is an international organization that focuses on security of software and software development. It has local chapters around the globe that meet regularly and offer various activities. In some cases, members may be actively working on one of the many OWASP projects. Other chapters may offer more of a social gathering in which people just openly discuss current topics in security.

If you find a local chapter, it would be good to reach out to the leaders to find out more about how their chapter is run and see if it is a good fit for you. OWASP groups may require you to be a registered member of OWASP, although you may be able to attend meetings without being a member. The leaders can tell you more about this as well.

ISACA

ISACA is an international organization that provides certifications to cybersecurity professionals. Its focus tends to be more on the practitioner in security, people who work in the security teams for their organizations that may not be in the cybersecurity business. ISACA has local chapters around the world. However, there is a cost to join ISACA, and becoming a member of a local chapter typically requires additional membership fees.

Reaching out to the leaders of a local chapter might enable you to find out more about the chapter and possibly attend a local meeting as a guest. This would give you a good opportunity to discover if it's the right fit for you before committing to an annual membership.

7.1.3 *Other meetups, conferences, and events*

Within the security community in particular, a lot of other opportunities exist to connect with professionals in the field. Security, possibly more than any other industry, offers various international, national, and local conferences that are often community led with the focus on developing skills and providing networking opportunities. In some cases, vendors of security products and services even host their own conferences. These are great opportunities, as discussed in chapter 5, for building your skills in security. However, they also can be a great opportunity to network with others, especially if you're more of the outgoing type.

Vendors, security groups, and other organizations will also host various one-time meetups or other events through which you could also expand your networks. These can be anything from seminars with guest speakers, to roundtable discussions, to even things like wine tastings or other activities. Keep on the lookout for these as well. You might find you have to sit through a bit of a sales pitch in many cases, but at the same time, you will get the chance to meet others in the industry and once again grow your own network.

7.1.4 *Making your network productive*

The key element in building your network is to have a plan and be intentional. As with most things, it is something that requires continued maintenance and attention, not something you create and forget. You have to be an active part of your network to keep it growing and thriving. If you go quiet in your network, those connections you established can slowly fade away.

Lay out a plan and set goals for developing your network. Make sure that any goals you set are realistic, but at the same time use them to keep you motivated. This means making a commitment to yourself to do the work and to hold yourself accountable to your goals.

As you build your network, look for ways to increase your involvement. Networking is a two-way street. While you're looking for your network to help you, also look for ways that you can help the people in your network. The more you get recognized as someone who is willing to assist others in achieving their goals, the more willing and excited the people in your network will be to help you out.

That said, don't be shy about asking for help. Look for those in your network whose opinion you trust to look over your resume. Reach out to those you might want to work with and see if they have any job openings. Even if they do not, they likely will be willing to help connect you with others who might be looking for someone just like you.

It seems like this should be obvious, but sometimes aspiring and even experienced professionals build a great network but then are afraid to use it. Remember why you spent all that time cultivating your network and then use it to your advantage.

7.2   *Exploring the role of mentorship*

In a general sense, a *mentor* is someone with experience whom you can work with as an advisor to help you grow in your career or make key decisions. Mentorship is a topic that comes up often in the security community, especially in discussions about helping new professionals enter the industry.

Mentoring is valuable, however, across all industries and at every level of career progression. Many top executives in high-profile organizations still work with mentors to develop and improve their skills. The important thing to remember with a mentorship is that it can and should be a mutually enjoyable relationship.

Finding a mentor can be a difficult proposition, and the way you go about finding one can vary. I've found that my most productive mentoring relationships formed organically. I didn't specifically set out to find a mentor, but instead over the course of time working with someone, discovered that they were a great mentor for me.

However, more formal ways to find a mentor are available. Some organizations have even created mentor-matching platforms to try to connect young or aspiring professionals with mentors in their chosen area of security. Multiple platforms exist, some of which charge a pretty heavy fee. One good free platform is Cyber Mentor DoJo (https://cybermentordojo.com). You can create an account and browse the list of mentors to find one that will work for you.

If you'd prefer a more casual method for finding a mentor, the hashtag #CyberMentorMonday on Twitter can be a great way to find or announce you're looking for a mentor. Regardless of how you find a potential mentor, it is important that you know what to look for to make sure they're the right fit for you.

7.2.1   *Qualities of a good mentor*

When looking for and selecting someone to ask to be a mentor, it's most important to ensure that they actually have the makeup of a good mentor. Some of the key characteristics might be what you expect, but others are perhaps not so obvious.

The first quality of a good mentor is *honesty*. You want someone who is able to deliver bad news as well as good news in an honest and up-front fashion. That is not to say that they need to be overly direct in their feedback, but you want them to call out areas for improvement. That is the whole point of having a mentor, after all. Unfortunately, some mentors are not well versed in the interpersonal skills necessary to deliver honest feedback in a productive way. As you look at your relationships with those in your networks, look for the ones who deliver the hard truth in a way that inspires you to take action.

Enthusiasm is another crucial quality for any good mentor. They need to have enthusiasm not only for cybersecurity but for sharing their knowledge and helping coach others. Plenty of people in the security community don't have a passion for this type of relationship. As you look for a mentor, keep an eye out for not only their passion in talking about cybersecurity topics, but also the energy with which they share their knowledge and ideas.

Any good mentor should also have their own *strong professional network*. A good mentor is someone who does not necessarily know all the things you want to learn but can connect you with others who have specific domain knowledge when it is needed. Your mentor should be able to point you in the right direction when you need help. Ideally, they should be able to make introductions for you when you need to dig deeper into a particular topic.

Consider how well connected to the industry someone seems to be before you ask them to be a mentor. However, also keep in mind that not every mentor you have needs to be in security. Having mentors that can help you grow as a professional, as a leader, or as an entrepreneur can be valuable, and they don't necessarily need to have cybersecurity knowledge to do that.

Finally, a good mentor candidate should be someone who *complements your skill set.* They should have the skills that you're looking to develop. If you want to become a better leader, you likely would be best served finding someone who has demonstrated success in progressively higher levels of leadership. If you are looking to grow your knowledge and skills around digital forensics, find someone who has spent considerable time working in that field. It is pretty straightforward, but it is not uncommon to see aspiring professionals expect things of their mentor that their mentor simply isn't qualified to deliver.

7.2.2 What to expect from a mentor

Going into any mentoring relationship, you should understand your expectations. That begins with making sure that your expectations are realistic. A mentor is there to guide and to advise; they are not there to teach you all the skills you need for a certain role. They should be able to direct you to learning resources and help you build a plan, but they won't be the ones teaching you those skills on a day-to-day basis.

You should absolutely expect that your mentor will work with you in a nonjudgmental and constructive fashion. They should be willing to listen and to be empathetic to the issues you are experiencing. Above all else, you should demand integrity from your mentor. If they are not ethical and do not behave true to the things that they preach, they are not going to be able to guide you appropriately through building your career.

Ultimately, a mentor provides five key elements: advice, planning, motivation, expertise, and support (figure 7.1). A mentor should be able to provide advice about situations you encounter, career decisions you're considering, and new ideas you are developing. Your mentor should also help with planning. This means setting career goals and a path for achieving those goals. A crucial aspect of any mentor is providing motivation. They should help foster your passion and keep you focused when your energy wanes. Expertise is also important. Your mentor should have demonstrated knowledge in the areas you're expecting them to help you with—whether that's technical cybersecurity skills, leadership skills, or simply professional awareness. Finally, a mentor should be there to support you. They should be an advocate for you and actively assist in building your network, locating potential roles and even helping recommend you for a new opportunity.

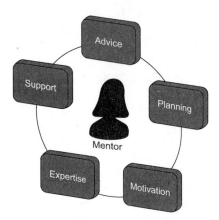

Figure 7.1   What you should expect from your mentor

In August 2021, I conducted a survey of over 1,500 current and aspiring cybersecurity professionals to ask about their views on mentorship. One question asked the participants to rank, on a scale of 1 to 5, the importance of each of the five mentoring elements in a mentoring relationship. The results seemed kind of boring at first, as each element scored similarly, as shown in figure 7.2.

Now, perhaps I could have improved the survey by using a wider scale, but as it turns out, some interesting data was still revealed. Despite the *average* scores being relatively similar across each of the five elements, the *individual* scores by respondent varied widely: each of the five received the lowest score of 1 while also receiving the highest score of 5, as well as responses between those values.

What does this mean for you? These results show just how individualistic a mentoring relationship can be. While a mentor should be able to provide any of the five areas, and all should be a part of that relationship, where you get the most value personally will be individual to you. Keep this in mind when you are considering whether someone might be a good mentor.

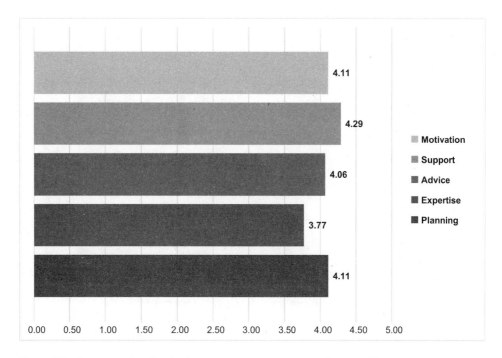

Legend:
- Motivation
- Support
- Advice
- Expertise
- Planning

Figure 7.2   Survey results showing importance of each element of mentorship

7.2.3   *What your mentor expects*

Just as it is important for you to understand your expectations of your mentor, it is important to have a clear idea of their expectations of you. As I said earlier, mentoring is a two-way-street that should be beneficial and enjoyable for both parties. Knowing your mentor's expectations of you in this relationship will help ensure that it is productive and motivational for both of you.

It is important to understand, first, why mentors even choose to be mentors. What is in it for them? In the mentorship survey, I also asked the respondents if they had ever served as a mentor themselves. Those indicating they had were then asked their motivation for becoming a mentor. Figure 7.3 shows the results.

As you can see, the overwhelming majority of those who have been mentors chose to do so as a way of giving back. This motivation is important to understand because it shows that for the majority of mentors, this relationship is about helping others more than any form of personal benefit. Even just the ability to have a rewarding experience wasn't as significant of a factor as a certain desire to help others grow. You can understand from this that mentors are there to help you grow. Their expectations of you will ultimately focus on your commitment to growing. So let's now analyze a few things you can do as a mentee to show your commitment.

A good mentor will expect you to have some level of career plan. You don't need to have a carefully laid-out progression, but identifying the goals you have and making

Why did you become a mentor?

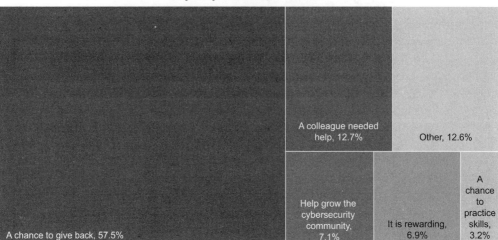

A colleague needed help, 12.7%

Other, 12.6%

Help grow the cybersecurity community, 7.1%

It is rewarding, 6.9%

A chance to practice skills, 3.2%

A chance to give back, 57.5%

Figure 7.3   Survey responses from mentors on why they became mentors

sure they're specific is critical if you want to have a productive relationship with your mentor. Going through the exercises in chapter 4 should arm you with a lot of great information that will help your mentor understand what you hope to achieve and then speak specifically to how you can get there. Remember, their job is to guide, but you need to know the ultimate objective.

Your mentor will also expect you to ask specific questions that address your needs for development. Coming to a mentoring session and saying something like, "Can you teach me how to hack?" or "I want to learn all of cybersecurity" is not going to allow them to advise you in any way. Again, a mentor is not an instructor who will teach you all the skills and techniques you need for a job. They are there to advise you on the best course of action to take in your specific situations and to guide you through your skills development by directing you to resources or answering specific questions about various topics.

Finally, and this is probably the most important of all, your mentor expects you to be respectful of their time. Understand that any mentor you find is most likely working a day job. They might also be mentoring other individuals, and they most assuredly have a personal life. So be careful of the demands you place on them for attention. You need to be willing to go off on your own and dig into topics after they have provided guidance or direction. Allow ample time for them to respond to communications from you. As with anyone, they may have days when they get done with work and the last thing they want to do is look at another email. Allow that space for them, just as you would expect someone to do that for you.

7.2.4   *How many mentors should a person have*

When it comes to finding a mentor, having more than one is OK and often necessary. On the flip side, there really is no one right answer for how many mentors a person should have. Mentoring relationships, as I said before, often come up organically. You might find that you have one mentor who is very technical and can help you with day-to-day skills growth, while a second mentor might help you develop more in terms of job search skills. This is perfectly normal and OK. Ultimately, you want people from your network to become mentors in ways that help complement your skills and help you grow.

All that said, it is possible to have too many. Mentorship is a valuable tool in your professional development toolbox, but it's only one tool, and you need to have many. It can be easy to fall into a trap of having regular mentoring sessions scheduled multiple times a week with a variety of mentors. At some point, you have to ask yourself, how much time am I spending talking with mentors versus actively working to grow my skills?

It is possible to become so reliant on various mentors that you fail to improve in any meaningful way yourself. So, pay attention to how your mentoring relationships are evolving. If you're not seeing value from certain relationships anymore, it might be time to consider shrinking the pool of mentors you draw upon.

7.3   *Managing the mentor relationship*

Great, so you think you found someone to be a mentor. You have connected with them on a level that has made you comfortable talking to them and you trust them to guide and advise you. Now it is time to recognize that every mentoring relationship has its own unique formula for success. There is no one right way to structure such a relationship, and both of you need to work together to form and grow the relationship.

7.3.1   *Forms of mentoring relationships*

Mentors can come in all shapes and sizes. As I discussed earlier, you may have different mentors who help you with different areas of your career. For instance, I regularly work with someone I trust to teach me how to navigate the corporate leadership aspects of my career. She is an executive in a large sales organization, and she doesn't have a lot of technical knowledge in cybersecurity. She guides me on progressing up the corporate ladder, but I would never expect her to help much with technical issues I'm having. Instead, I work with other mentors who have great breadth of knowledge across different security technologies. So, as you develop mentorships, know that you'll likely want to follow a similar model.

Some mentors might even focus on helping you build your network more than they help with building any specific career skills. Some people in my network, whom I consider to be mentors, are the first people I turn to when I want to get connected with someone in the industry for some reason. I know they have large professional networks and they usually can find a way to get me an introduction to the person I'm trying to connect to. So be open to what mentorship might mean; it is not always a formal career coaching relationship and might be far simpler.

7.3.2 *Structure of a mentoring relationship*

How you choose to work with your mentor may vary based on the type of relationship, the way it developed, and your goals. Some mentorships are structured, with regular meetings, goals, and action items. Others can be more ad hoc. It is all about finding the format that works best for both of you and makes your interactions enjoyable and valuable for all parties involved.

Regular check-ins are a great way to set up a more formal mentorship. This is often the type of structure you would choose if you found a mentor through a matching service or through a colleague in your network. Having weekly, biweekly, or monthly meetings with your mentor can be valuable. This allows you to plan for topics that you want to discuss and to provide updates on how you are progressing. It also ensures a level of accountability that can be important as you grow in your career path.

However, other mentorships may be more ad hoc—in particular, those relationships that grow from more organic interactions. If you didn't set out specifically to find a mentor, this could be a person you work with on an as-needed basis. You might schedule a call when you want to talk through a specific situation or even possibly send questions to via email. This type of structure can be just as valuable as the more formal periodic meetings, and it also allows for less time commitment from your mentor. However you choose to interact, though, it is important that you and your mentor discuss that methodology and agree on the ground rules for managing it.

When you and your mentor do meet, the format of those meetings can vary as well; see figure 7.4. Some of the best mentoring relationships I've had are with people I could call and just schedule a time to grab coffee together. We could catch up and talk about the issues we were facing or the successes we were experiencing. It's a great way to learn to see your mentor in a more human light.

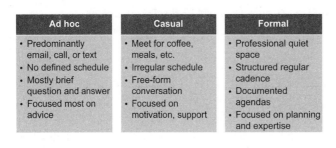

Figure 7.4 Examples of varying structures for a mentoring relationship

However, having structured conference calls with a formal agenda and even action items at the end of the meeting can also be rewarding. The goal here, as I've said throughout this section, is that you find an arrangement that works well for both you and your mentor.

The format that you choose is, once again, a personal decision. Choose the format that not only is most effective for you, but also that your mentor can commit to. In fact, this could even shape your decision on whether a particular person would be a good mentor for you. If you are looking for a formal mentoring experience, but they cannot commit to that, perhaps you want to continue looking elsewhere for someone whose current workloads permit that type of relationship. On the mentoring survey, respondents were asked the type of mentoring relationship they found most effective. I must admit, I was a bit surprised by the results (figure 7.5).

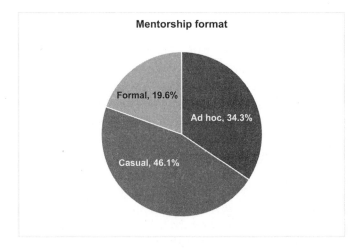

Figure 7.5 Survey results showing the preferences for the three types of mentoring relationships

The overwhelming majority of people do not seek out mentorships of a more formal format. This was surprising to me because a lot of the mentorship discussions I see and hear refer to that format. So clearly it is important to recognize that more casual or even ad hoc arrangements need to be considered when you connect with your mentor. These arrangements can offer a more flexible approach and feel potentially less threatening. Be honest about which format will suit you best and remember that over time you can always change the structure if things aren't working out as you anticipated or your needs simply change.

One last note on the format of your mentoring relationship. Throughout this chapter, I've discussed the mentoring relationship as a one-on-one situation. After conducting this survey, however, I got feedback from individuals indicating that group settings can also be effective. In this case, a single mentor may meet with multiple mentees at the same time. This type of arrangement is often found in the form of study groups or other similar group settings where the interactions with a primary mentor is only one part of the larger dynamic. Having peers in a group setting like this for some can be less threatening. It also provides additional perspectives and experiences that could be helpful, especially when focusing more on technical expertise and skills growth. Therefore, it cannot be stressed enough how important it is to keep an open mind not only about who could be a good mentor but also how you might structure such a relationship.

7.3.3   *Ending a mentor relationship*

Sometimes you will reach the point where your relationship with a mentor has run its course and it is time to end the relationship. Knowing when this is the case is not always easy, and worse, that can be difficult to admit to yourself. You have spent a lot of time working with this person, and they have helped you through many things. You may end up feeling a certain sense of loyalty to your mentor, and that can make it hard to exit the mentoring relationship.

It is important to regularly examine the value you are getting from your mentor. Do you find sessions with them to be arduous? Do they seem disconnected or even frustrated when you talk? Are you learning new skills or getting valuable direction to grow your career path still? You need to be objective and identify when the answers to these questions indicate that the relationship has simply hit the end. Like any relationship in your life, they don't always last forever. You can still be colleagues and even friends, but that mentorship side may just be at an end.

Ultimately, ending the mentoring relationship is not something to fear or feel guilty about. But as with ending any relationship, you want to do it gracefully and with respect. If your arrangement with your mentor has been more ad hoc, it could be as simple as just not continuing to schedule sessions. However, the real challenge is when you have a more formal arrangement. Just remember, sharing that you are no longer getting the value that you once were should carry no shame or guilt. Thank them, be gracious, but also be honest, just as you would expect from them. Even if they are not formally your mentor anymore, they still are a valuable part of your professional network, and you don't want to damage that.

7.4   *Building those relationships*

Now that you understand the roles that a strong network and mentors can play in your development of your cybersecurity career, it's time to get started. Set up those social media accounts if you have not already. Begin engaging with industry peers and those whose work you find interesting or impressive. Start searching for the various security communities that may be available near where you live.

Of course, start looking at your peers, coworkers, and colleagues to see if there are a few that you could trust to act as a mentor. Perhaps start looking into online resources to find that first mentor. It's never too soon to begin that search or to start adding to your professional network.

Summary

- Social media, industry groups, and local events can be great tools for building a professional network.
- Your professional network should be deliberately cultivated and developed.
- You should look for certain crucial characteristics anytime you are considering someone as a potential mentor. These include honesty, enthusiasm, a strong network, and skills complementary to your own.

- Mentoring relationships can take on various structures, but ultimately both parties should get enjoyment and value from the situation.
- Ending a mentoring relationship can be necessary when the connection has run its course or doesn't work out, but you need to move on respectfully and gracefully.

The threat of
impostor syndrome

This chapter covers

- Understanding what impostor syndrome is, why it's important, and who experiences it
- Examining yourself to identify the causes of your impostor syndrome
- Avoiding or overcoming impostor syndrome
- Objectively recognizing and celebrating your achievements

Until now, we've focused the bulk of our attention on preparing for and landing that first job in cybersecurity. Now it's time to start looking at the forces that could work against you after you've landed that first role. Reading various studies from the medical, psychological, and career development industries, you'll find that an estimated 70%–85% of people experience a phenomenon known as *impostor syndrome*.

As we discussed in chapter 2, the cybersecurity industry suffers from a lingering rockstar culture. With so many large events, well-recognized personalities, and wide ranges of expertise, impostor syndrome is particularly common in this field as well. All right, you ask, what is impostor syndrome, and why should I care? Well, it is time to answer some of those questions and more.

8.1 Defining impostor syndrome

A lot of discussion occurs within the security community, at conferences and on social media, about people experiencing impostor syndrome and its negative impact. It's a real issue that threatens to derail the career aspirations of many in this industry. To combat this phenomenon, we must first understand what it is and why it is important, and recognize whom it can impact.

8.1.1 What is impostor syndrome?

In the simplest of terms, *impostor syndrome* is the belief that your accomplishments aren't as valuable or impressive as others perceive them to be. It's a sense that you are not as qualified to do the things you are doing as others believe you to be. This is an internalized state of mind. Many people who experience impostor syndrome describe a fear of being discovered as a fraud. In the professional sense, people look back on their accomplishments and qualifications and tend to minimize their importance or relevance.

Impostor syndrome within cybersecurity can also come down to questioning your breadth of knowledge in the field. Figure 8.1 illustrates how we often perceive our own knowledge in comparison to those around us working in the industry.

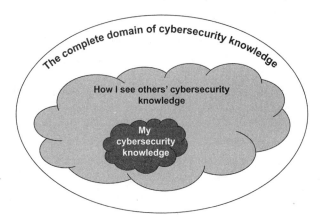

Figure 8.1 A graphical representation of how impostor syndrome colors our view of our own knowledge in comparison to others

As you can see, it is not uncommon for individuals in our industry to believe that everyone around them has a far greater knowledge of cybersecurity than their own. This sometimes results from the human tendency to collectively combine the knowledge we see others share and then attribute that collective capability to everyone else. We then recognize that our knowledge overlaps—but is only a subset of—that larger collective.

However, this is not true, nor is it at all realistic. In reality, the comparisons of our own knowledge to those around us are much more like figure 8.2.

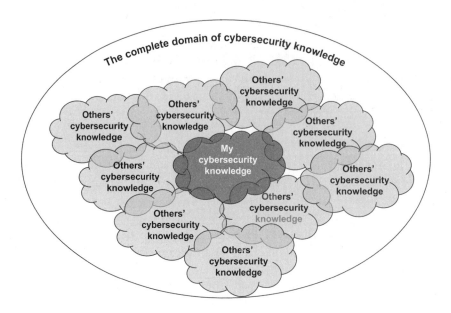

Figure 8.2   A graphical representation of how cybersecurity knowledge is distributed among members of the industry

The reality is that everyone we see around us has their own subset of knowledge, but no one knows everything there is to know about cybersecurity. They could not; it would not be possible.

Additionally, as you can see in figure 8.2, each person's knowledge overlaps with others', including our own. At the same time, not everyone's knowledge overlaps with everyone else's. Additionally, clear differences exist between each person's base of knowledge, and each person has a unique knowledge set they bring to the table.

While the overlaps in knowledge create depth in our capabilities, it's the unique areas that drive the myriad of perspectives that make cybersecurity stronger as a whole. This is an important concept to understand as you work to combat your own experience of impostor syndrome.

8.1.2   Why do we care about impostor syndrome?

Well, this is all great information, but what does this have to do with career success in cybersecurity? Impostor syndrome is a powerful force that can have a detrimental effect on your career progression. The impacts of impostor syndrome are far-reaching and diverse. But in the end, the toll they take on us from both a motivation and mental health perspective cannot be understated.

Perhaps one of the more obvious damaging effects is that the fear of being discovered a fraud holds us back from risk taking. When a person undervalues their accomplishments or minimizes their legitimacy, a gap is created in self-confidence.

If an opportunity to advance their career should arise, they may choose to forego that opportunity, feeling that they are not qualified and that if they pursue it, they will be found out. This is a particularly common impact that is well-documented in various studies of the phenomenon. It is important to recognize that impostor syndrome affects us in this way so that we can take steps to combat it rather than let it hold us back.

Impostor syndrome can be detrimental to mental health as well. Imagine the scenario in figure 8.1: believing that all your peers have a tremendous wealth of knowledge that you somehow lack can generate crushing levels of anxiety. Consider how overwhelming it would feel if you believed you had to gain all that knowledge to make your cloud as big as everyone else's. The perceived task would be monumental and almost unachievable. The resulting impact can damage not only your desire to pursue future opportunities but also lead to a disdain for your current role. In the worst-case scenario, it could lead to you abandoning your career path in security altogether.

Another mental health impact from impostor syndrome can be self-loathing. If we see ourselves as inescapably underqualified, we may begin to view that as a flaw in ourselves. Indeed, this feeling is often expressed in studies and even more informal discussions on social media. It can become a vicious cycle, in which our feelings of inadequacy lead us to demean ourselves, which leads to further feelings of inadequacy, and so the cycle continues.

Again, this unfortunate perspective can lead to abandoning a career path altogether if left unchecked. As a result, it is important that if you are to be successful in your pursuits, you recognize and actively work to combat these influences.

8.1.3   Who experiences impostor syndrome?

If you look at articles and published papers on the topic of impostor syndrome, you'll quickly discover that this phenomenon is experienced almost universally. Regardless of people's career stature, knowledge levels, or public image, almost everyone reports having some level of experience with impostor syndrome. As stated earlier, many studies suggest anywhere from 70%–85% of professionals report experiencing these feelings.

Consequently recognizing that impostor syndrome can and does affect us all is crucial. Whether you are just starting your career or are well established, the feelings of doubt and inadequacy can be a challenge. Anecdotally, by observing discussions in the cybersecurity community, even those who are recognized as leaders will report they experience the feelings and effects of impostor syndrome. I can report at a personal level that I have had to battle this perception throughout the almost two decades I have spent working in cybersecurity.

Let me share a little more of my personal experience. I began working in cybersecurity as a penetration tester in 2006. I had never considered hacking as a career path but now found myself as part of a security testing team within a large financial services company. Within a year, I was the lead for that team. By 2010, thanks in part to an acquisition, I was now managing not only that team, but also the complete vulnerability management program for this financial services organization of 35,000 employees, which appeared in the top 200 of *Fortune* magazine's list of the top 500 companies at the time. I was barely over 30, yet I had a prominent role in the operational security of this global entity.

That sure sounds like an impressive career feat, doesn't it? Yet for many years after I left that company, I refused to look at it or talk about it in that way. Instead, I chose to minimize it.

I wrote it off as dumb luck and not really that monumental of an achievement that I had any right to celebrate or brag about. Even on my resume, I downplayed the extent of my responsibilities in that position. I rarely talked about my regularly reporting on our security posture to executives. I didn't describe working with the executives to secure funding for additional security initiatives and overcome the challenges of limited budgets. I was even bashful about disclosing the extent to which I had to work across the organization to ensure that vulnerabilities were remediated.

I looked at the help that I got from my previous managers, including the one who brought me into security in the first place, as a sign that my accomplishments were somehow less. I regarded the lucky break of an acquisition that thrust me into that level of responsibility as somehow diminishing the value of what I had undertaken in that role.

To put things in perspective, everyone achieves success through a combination of their own skills, their own risk taking, assistance they receive from others along the way, and in most cases, some measure of fortunate circumstances. The value of our accomplishments isn't measured by those factors. Instead, it is how we choose to react to those opportunities that leads to our success.

It took me a long time to come to terms with that, and it did hold me back from chasing higher-level opportunities earlier in my career. Recognizing that many go through this and that you can expect to as well will help you better manage and overcome those forces of doubt. We will discuss this concept further later in this chapter.

8.2   *Understanding the causes of impostor syndrome*

Many academic studies and media articles have been published on impostor syndrome, what causes it, and who experiences it. Rather than regurgitate the results of those studies, I will focus on sharing what I have learned about causes of impostor syndrome from my own experiences and my discussions of the topic with others. A quick web search will yield plenty of results if you wish to dig deeper into formal academic studies, but in this chapter, I would like to bring more of the cybersecurity context to bear.

8.2.1   *Perfectionism*

One of the most common causes of impostor syndrome that I and others experience is the level of exceptionally difficult or even impossible expectations we place upon ourselves. Several of my colleagues have shared that they are not the type to go into any new venture only half-committed. I too share this personality trait. When I take on a new challenge, I strive to ensure that I'll be competent and well educated in that subject.

By way of example, many years ago when I chose photography as a hobby, I didn't just buy a camera and go out and start taking photos. Instead, I spent weeks studying the technical aspects of how cameras work. I learned about the relationship between shutter speed, aperture, ISO rating, and such. I participated on multiple photography forums and learned the language, which also drove me to understand additional concepts like f-stops and exposure compensation. I also didn't stop with technical details; I studied the more artistic aspects of photography as well, including how changes in lighting, composition, color balance, and framing can impact the mood captured in a photo. Finally, I researched many cameras to ensure that the one I eventually purchased would fit me well and be capable of supporting the various creative aspects I wanted to achieve.

I held myself to high standards in launching my photography hobby, and these self-imposed expectations are like those many have described to me in terms of their impostor syndrome. When we set such high expectations for ourselves, they can become imposing or even impossible to achieve. We set impossibly perfect goals, and when we fail to achieve them, we let those failures convince us that we simply aren't good at that activity. Instead of leaving room to grow and learn, our perfectionism drives us to be unfair and overly demanding. This is a core cause of impostor syndrome.

8.2.2 *Industry expectations*

It is not just our own expectations that can put undue pressure on us and spur feelings of being an impostor. The cybersecurity industry itself can often place similar pressures on us. As we discussed in chapter 3, this industry overall expects practitioners to have a high degree of capability from the start. A great deal of pressure is placed on cybersecurity professionals to make things perfectly secure. Given the criticality of defending our digital way of life, the industry rarely makes room for practitioners or organizations to fall short of perfection.

Unfortunately, this sets us up for failure. In reality, mistakes will occur in technology. An odd dichotomy exists in that security practitioners recognize that these goals are unrealistic, yet we continue to try to achieve them. It is often stated that attackers will inevitably find a weakness that we've missed in our systems and exploit them. In the best-case scenario, we may identify and respond quickly and successfully to attacks and breaches, but to think we can be unhackable is unrealistic. The industry simply does not measure success according to reasonable and achievable metrics.

As a result, when these situations do occur, it is all too easy for security professionals to take those events personally. It is not uncommon for a security professional to look at a breach as a sign of their own failures or incompetence rather than acknowledge that technology is infinitely dynamic and difficult to secure as a result. This can be damaging to self-confidence and lead to feelings of hopelessness and anxiety that further impact our ability to perform our daily duties.

It is important for our career as well as our personal mental health that we view these types of challenges differently. Cybersecurity professionals need to regard them not as signs of our lack of qualification or competence but rather experiences that we can use to learn and to grow. It is a sign of true professional maturity to understand that learning and growth come from failures. If we were always successful in everything we try to do, there would be no challenge and hence no success to celebrate.

8.2.3   *Comparison to others*

As you saw in figures 8.1 and 8.2, a great deal of impostor syndrome comes from comparing ourselves to others. As you enter the field, you may establish relationships with other new cybersecurity professionals who are launching their careers around the same time. These relationships are important, and they can serve you well as you grow and advance in your career. However, they can also present challenges regarding feelings of inadequacy. Specifically, as you watch your peers progress in their careers, it is not uncommon to become focused on their successes and believe that they are progressing faster than you.

This type of comparison is inherently flawed. The reality is, typically, that they indeed are experiencing successes you have not, but you have also likely experienced successes that they have not. In this situation, you fail to recognize your own value and focus instead on what you have not yet accomplished. Even worse, those accomplishments may not have even been things you were targeting. Yet, seeing someone else achieve them still makes you feel like a lesser success.

The problem of comparison can also come into play when we look at the rockstar culture of the security community. You have probably already discovered the names of a few highly recognized and accomplished individuals in the cybersecurity community. You may admire their accomplishments and their skills. Perhaps they quickly ascended through various career milestones that you hope to reach as well. This is good and healthy if it motivates you to grow and achieve great things in your own journey. However, it becomes a problem when it leads to deep feelings of anxiety or pressure that become overwhelming.

It is common to look at all they have done and feel hopeless, as if you will never achieve their level of skill or success or that you are falling behind somehow. But believing that is true would be unfortunate. Remember that what you usually see of others in the public space is their best self. For most, that is the only part of what they allow you to see. You may not get to see the struggles and failures that they have dealt with and continue to deal with along their path. It can be incredibly surprising sometimes when you discover just how much they feel like they have not accomplished.

It is, therefore, critical to acknowledge that each of us is on our own journey and that accurately comparing one person's success with that of another is ultimately impossible. Career successes can come at the expense of successes in other aspects of a person's life. Hidden privileges or advantages may have helped one person reach a pinnacle, while others do not benefit from those same privileges and have to work through additional challenges as a result. This is why measures of success should instead be a personal thing. The way we identify our own milestones should take into context the many personal facets of our journey that impact our career and professional successes.

8.2.4   *Lack of representation*

In chapter 1, I discussed the need for diversity in the cybersecurity community. A large diversity gap exists in this space. Women, people of color, and other demographic groups are underrepresented. As a result, finding role models who have achieved the level of success that we aspire to and that we can also identify with on a personal level may be difficult.

For instance, for a Black person who aspires to achieve a high-level executive role, finding a Black role model in a leadership position to serve as a mentor or inspiration can be difficult. In the 2017 "Global Information Security Workforce Study" released by (ISC).² and Frost & Sullivan and others (http://mng.bz/06wm), only 9% of cybersecurity professionals were African American or Black. Among those, only 23% were in leadership positions. Logic would then also dictate that subsequently an even smaller subset would be found in executive-level positions.

We may want to believe that people's ethnicity, race, gender, and sexuality don't matter when it comes to identifying role models. However, looking at the situation more pragmatically, when members of those groups fail to see their demographics represented in their career field, it can lead to a feeling of not belonging. This feeling further exacerbates the already difficult feelings that result from impostor syndrome or can trigger feelings of being out of place in people who would otherwise not have such doubts. Given that much work remains to be done in terms of diversity in this community, it is important that you, especially if a member of an underrepresented group, look within and acknowledge that we are not only allowed but encouraged to blaze a new trail if necessary.

8.2.5   *Diminishing accomplishments*

I discussed earlier my personal experience of belittling my own accomplishments. Because I had received help from others and had also been the beneficiary of many serendipitous opportunities, I minimized the value of my career achievements. As it turns out, this is a common mindset among highly driven individuals, especially in cybersecurity. A flawed impression exists that those who've demonstrated the highest levels of achievements got there of their own accord, without outside influence. The reality is quite the opposite, however.

Success is rarely, if ever, achieved by working and conquering challenges alone. Athletes work with coaches, trainers, nutritionists, and others. They leverage this assistance to ensure they are the most fit they can be for their sport. They learn from others and find opportunities through those they work with. As another example, musicians spend countless hours playing with and learning from other musicians.

They might learn music theory or gain inspiration from others. Very few of the songs you listen to are written, composed, and produced by a single person.

In addition, neither athletes nor musicians, nor any other highly successful and recognizable person gets to their position without some measure of fortunate circumstances. Perhaps it is the lucky break that a professional scout happened to show up at a college baseball player's biggest game of the season. Maybe that recording executive just happened to be in the club the night that a particular band played there.

The same is true for your career in cybersecurity. You will have mentors, coworkers, and others who help you develop your knowledge, your skills, and even help you find professional opportunities. You will undoubtedly experience those lucky breaks that present you with an opportunity to take on a new challenge or leap into a new role. Ultimately, you'll find that success comes when you use those influences and those opportunities to your advantage. Taking risks and chasing the next big thing is often what makes the difference between a quick career progression and one that lumbers along.

Therefore, you would be remiss to think that because you or anyone else experienced such influences, those accomplishments are somehow less valuable or impressive. To drive your career progression forward, you need to see these events more objectively and be willing to give yourself credit for the way you leveraged those influences to grow and advance. That is what makes those achievements something you can be proud of.

8.3   *Overcoming impostor syndrome*

Now that you understand what impostor syndrome is and the detrimental effects it can have on your career journey, we can start to discuss what can be done to limit those impacts and possibly even avoid the phenomenon altogether. Since you now understand that everyone is susceptible to these feelings, you know that you don't need to feel ashamed or broken if you struggle with them too.

Not everyone's experience of impostor syndrome is the same, so the ways to avoid or overcome the challenges will be unique to our own journeys as well. There is no one right way to go about it, and in reality, the methods we find successful can change over time as well.

What will not change is that impostor syndrome will not go away on its own. It requires acknowledging that it exists and that it could be impacting you. Developing ways to see yourself, your journey, your struggles, and your achievements in a different light will give you the tools you need to defeat this nemesis. So, let's look at a few tactics you can leverage in your career journey to ensure that impostor syndrome does not hold you back from achieving all the success you hope to realize.

8.3.1   *Avoid competition*

As you have seen throughout this discussion of impostor syndrome, many of the causes and challenges come from the expectations we place on ourselves and comparing our journey to that of others. In particular, these can lead to a sense of competition with those around us. We want to achieve that milestone faster than someone else got to it. We want to have more credentials applied to us than that peer we see in the industry. We want to know more about a given topic than someone else. The list of expectations and competitive motivators goes on and on and on.

Unfortunately, the cybersecurity community sometimes contributes to this natural tendency for competition. We regularly organize events like CTF challenges, in which individuals or teams compete to hack various systems. It is assumed that the most skilled will accomplish those tasks the quickest and win the prize. These competitions aren't inherently bad or unhealthy, but if we use them as a measure of our own skills rather than simply an opportunity to learn and grow in a fun competitive environment, they can be toxic. If you choose to partake in these activities, try to see them for what they are worth. Just as sports participants can become overly competitive, the same can happen in these spaces. Keep it fun and understand that just because you or your team doesn't win doesn't mean that you're somehow less valuable to the community.

Competition in cybersecurity can also occur in the discovery of security vulnerabilities. Those new flaws we find, which we refer to as *zero-days*, are often leveraged to establish someone as a skilled researcher. These zero-day vulnerabilities are reported typically via the CVE database. So, you may encounter people who measure a hacker or researcher's skills by the number of CVEs they have reported.

This is dangerous for a lot of reasons. First, discovering a zero-day vulnerability can be valuable but shouldn't be seen as having greater value than finding long-recognized vulnerabilities in a system or application. Some may argue that the latter is more valuable in terms of securing systems. Additionally, many hackers and researchers (myself included) work for organizations testing and securing their systems. Vulnerabilities discovered in their systems or applications may be zero-day vulnerabilities, but because they were found in software developed by the organization (rather than sold commercially), the vulnerabilities aren't reported to the CVE database.

The point of all of this is that you will be tempted to compare and compete with your peers in many ways throughout the cybersecurity community. Competition should be approached carefully, however. It should be leveraged for fun or as a motivator to continue to grow your expertise and your career. It crosses the line and becomes a damaging influence when it is no longer enjoyable, and instead you start using it to measure your own value. Be cautious about competition and know that just because you don't win that prize, get that accolade, or see your name credited on something does not mean you aren't contributing.

8.3.2   *Set goals and define what success means to you*

Related to the topic of competitiveness and comparison is the idea of how we measure our success. The reason that things like CTFs and CVEs become measures of success for some is that they are easily demonstrated to others as success indicators. In fact, you can fall into many traps in terms of your career journey and impostor syndrome if you are defining success by the way others will recognize it or become aware of it. This is a toxic way of defining success. Just as our journeys are unique to each of us, so too should be our measures of success. Defining achievements based on the perceptions of others versus what is meaningful to us subjects us to a constantly changing and ever dynamic set of metrics that will be impossible to achieve.

If your motivations for attaining a certain position are focused on the way you will be perceived by others, that is a tenuous and ultimately dangerous goal. The old adage that you can't please everyone comes into play here. No matter how big your success, there will always be those who will criticize and doubt you. Sadly, that's just a part of human nature, especially in the cybersecurity community.

Additionally, because of the dynamic space in which we operate, setting your goals based on others' perceptions of an impressive achievement today leaves you with no guarantee that when you realize that goal, it will still be seen as impressive. Pursuing success for the sake of fame, acclaim, glory, or whatnot leaves you with little room to shift and change. Anything short of being among a very small percentage of people suddenly means failure. That is not a healthy perspective.

Instead, it is important to set goals that are important to you. Consider the meaning that those goals have for you. Why is that something you want to achieve? Do you want to be an executive-level leader? Why? If it is because you want to be respected by others as a top expert, you may be disappointed someday when you become a CISO. Look around on social media or in the news. Pay attention to how often CISOs of organizations or in general are treated with disdain. Setting that goal based on how others will perceive you is not going to work out well. Now, instead, if you want that role because you relish the challenge of bearing the responsibility for the cybersecurity posture of an organization, or because you feel like you are well suited to defining strategy and interacting with other executives, well then, that goal might be the right one for you.

But as you will see in chapter 9, goal setting needs to be more than just identifying a big objective at the end of the road. We need to also set shorter-term goals that help us get there along the way. If your only goal might take 7, 10, or more years to reach, falling into the trap of devaluing what you have done along the way to get there becomes too easy. Accordingly, keep reading, as we will discuss later how to set attainable goals that keep you motived and progressing toward your eventual objective.

8.3.3 *Turn to colleagues and peers*

No, you don't want your goals to be based in what others think of you or how impressive they find your accomplishments. However, that does not mean that your peers and your colleagues can't provide effective support in combating the feelings of impostor syndrome.

One of the unfortunate symptoms of impostor syndrome is that professionals tend to have a hard time accepting compliments. Think about it. How do you react when someone tells you what a great job you did or how impressed they are by something you were able to pull off? Do you thank them and enjoy the feeling of having someone share in your success? Or do you let the feelings of flattery cause you to deny the value of what you were able to get done?

In our society, we are conditioned against bragging. However, that conditioning can go too far and cause individuals to be excessively humble when it comes to admitting their own worth. Fear of being seen as conceited or regarded as a braggard will cause individuals to downplay compliments. Additionally, to assist in that denial, it is human nature to assume that the compliment is biased or that "they're just being nice" rather than accept the genuineness of the sentiment. When it comes to impostor syndrome, you can flip this narrative on its head and use it as a form of validation.

When you detect those feelings of being an impostor, that is a good time to reach out to someone you respect and share those feelings. You likely will find that they feel similarly. It might even be jarring to you, as the things they feel worst about might be what you respect most about them. Do not shy away from having those conversations. Chances are they will share with you that many of your concerns are related to aspects of your journey that they most admire. Keep your mind open and be willing to accept that their comments are genuine. Just because you initiated a conversation on the topic does not mean that what they tell you is biased.

Your mentors can also serve a valuable role here. They might have additional context of similar feelings they have experienced. They could potentially share with you tools that they used to overcome the feelings. Find the solutions that connect with you and leverage them in your own journey. This is why gaining that personal perspective of your mentor as we discussed in chapter 7 is so critical. Make that relationship work for you and help you to see that you are not a fraud, you are not underqualified, and you do ultimately deserve the accolades you receive.

8.3.4 *Be a resource to others*

You are early in your cybersecurity career; I totally understand that. As a result, you may feel like you are not in a position to help others in their career. You would be wrong. As you take the first steps in launching a career in this field, you already have the ability to serve as inspiration for others who are considering a similar career path. As you continue in your journey, openly sharing your experiences can help others grow and advance as well.

A wonderful side effect of this effort is that it can help you conquer your feelings of impostor syndrome. The experience of helping raise others can be affirming in multiple ways that directly speak against the forces of impostor syndrome. Consider first the discussions you will have with those who seek you as a source of inspiration. Something about what you have accomplished speaks to them or is something they can identify with. As you share your experience and they react, you will see how various elements of your journey that seemed insignificant to you are actually quite valuable. This affirmation of your success will help to address that feeling that you have not yet achieved anything that comes from not being at your goal yet. When talking to someone who is inspired by you, it is also easier to accept the authenticity of their feelings. Listen to what they say, share what you know, and you will understand just how far you have come along this path.

Being a resource to others is also a chance for you to demonstrate your knowledge in various technical aspects of cybersecurity. Skills and techniques that you have learned and taken for granted will suddenly be called upon. As you begin sharing that knowledge with others, you will begin to realize just how much you have grown. It can serve as a reminder of where you once were in your own journey and the degree to which you have developed your own skills and knowledge. Let it be an opportunity for self-discovery as much as teaching and lifting up other members of the community.

Finally, as you help to build others up, no matter what stage of your own career you are in, that experience will help you along the path to your own goals. If your long-term goals are to achieve a position of leadership, the experience of helping others grow will equip you with effective leadership skills. If your goals are based more on technical acumen, imagine the benefit of having to potentially do some research of your own in order to help someone else with a problem they are experiencing.

This is where the adage of a rising tide raises all ships comes from. The more you do to help others grow, the more you will grow yourself. In that growth, you will achieve goals faster, which in turn will help you to ward off those feelings of impostor syndrome.

8.3.5   *Acknowledge and celebrate your achievements*

One of the toughest aspects in dealing with impostor syndrome can be looking at our past accomplishments in an objective light. As I described earlier in my own story, I found it difficult to admit to myself and others the levels of achievement I had reached at fairly early points in my career. I was 19 years old, had not even graduated college yet, when I got my first full-time job as a programmer. That would be a big deal for many folks, but I just looked at it as I got lucky because it was during the "dot com" era and programmers were in high demand at that time.

I've discussed at length the challenges of admitting our successes and recognizing that receiving help and getting lucky breaks are not just OK, but necessary components in achieving success in any field or pursuit. As you progress on your journey, it is important to acknowledge those achievements and allow yourself to celebrate them and be proud of them.

You might be asking yourself right now, how do I make sure I recognize those achievements? Let me share with you a simple exercise that you can perform to help you recognize the value of what you have done. Now, presumably, you have not yet launched your cybersecurity career; that is OK. We will just use your overall journey to date. You can work with your academic experiences, your work experiences in other fields, or even a hobby. Any of these will work for this exercise. Choose one or even a couple, grab a piece of paper, and let's begin:

1 Think back on the experiences you had. How did you learn how to do those things? Start listing any educational steps you took, any people you worked with who helped train you or shared knowledge with you, and describe any lucky breaks you experienced along the way.

2 Separate from that list, describe the most enjoyable facets of that experience for you. It could be an aspect of the job that you enjoyed (even if overall you hated the job), it could be that you achieved a certain success or milestone, perhaps even just a level of personal satisfaction that it brought to you.

3 Look at your first list and the subsequent description of enjoyable aspects and connect the two. How did those educational steps, the people who helped you, or the lucky breaks contribute to those enjoyable aspects of your experience? List them.

4 Considering those influences, describe how you used those outside influences. How did you apply those in a way that helped achieve those enjoyable moments? Focus on your actions. What did you do with that information or that opportunity that led to your personal enjoyment?

5 Take those descriptions of your actions and rewrite them as if you were putting them on a resume. It does not matter whether they are job related. Consider how you would tell others about them to paint yourself in an impressive light. Do not exaggerate or lie; just simply tell the story in a way that you believe would help others appreciate what you did.

6 Take a step back and consider what you have written objectively. Try to forget for a moment that you wrote these about you and think instead about how you would feel if you saw those items on someone else's resume or biography. Now, remind yourself that you wrote these things about you. There were no lies or exaggerations; those accomplishments are yours to be proud of.

Keep what you have written. This will be a good reminder of this exercise so you can repeat it as you progress along your career journey. Use these steps to appreciate in a more objective way just how much you have been able to achieve as you grow and develop in your cybersecurity career. In this way, you will be able to tackle and overcome that monster of impostor syndrome that keeps trying to tell you that you don't belong or that you are not enough.

Summary

- Impostor syndrome is the undervaluing of our own achievements and feeling like we don't belong.
- Impostor syndrome can keep us from progressing in our careers and can be experienced by anyone at any stage of their journey.
- Impostor syndrome can be caused by perfectionism, outside expectations, comparing ourselves to others, a lack of role models we identify with, and devaluing our own accomplishments.
- You can defeat impostor syndrome by avoiding competition, setting personal goals, getting support from colleagues and peers, and objectively recognizing and celebrating your achievements.

Achieving success

9

This chapter covers

- Recognizing and overcoming challenges to career success in cybersecurity
- Setting long-term goals and building a progressive strategy to achieve them
- Pivoting from one area of cybersecurity to another
- Taking everything you've learned and putting it into motion

In chapter 8, we transitioned from discussing how to land your first role in cybersecurity to addressing long-term success. We focused on one of the key challenges that cybersecurity professionals face from the earliest stages of their career journey throughout their development and beyond: impostor syndrome. However, that is only one challenge that threatens the success of your career. Because I want this guide to help enable your career for long-term success, it is important to discuss those other challenges in detail and identify ways of overcoming them.

Launching a career, whether as your first path directly out of school or as a change in your journey, is a risky undertaking. Having a strategy for building that career gives you steps to follow and clear intermediary objectives that you can focus on and measure to maintain your focus. The strategy you set forth at the start of your journey will no doubt change over time. Chapter 2 covered the various disciplines that make up cybersecurity. The specialty you choose today will probably not be your focus for the rest of your career. Still, the planning you do now will serve as your guide for years to come.

That propensity for your path to be dynamic and shift over time makes cybersecurity a particularly attractive career choice. As a result of the many varying but related disciplines, cybersecurity offers a particularly easy path for pivoting from one area of specialization to another. Still, it can be daunting to make such a change.

To help equip you with the tools necessary to make this switch effective, this chapter discusses *pivoting*. You have come this far in this guide; now it is time to put all the knowledge you have gained into motion.

9.1 Overcoming challenges in cybersecurity careers

Impostor syndrome was discussed at length in chapter 8 because it is such a universally encountered difficulty and one you will likely do battle with at various stages in your professional journey. However, it is not the only difficulty you may face that threatens to derail your plans for a long and successful career in cybersecurity.

Positioning yourself for that long-term vision requires anticipating and navigating those problem areas that could arise. While they are not unique to cybersecurity, the areas discussed here do exhibit themselves in unique ways in this domain. They are common topics in cybersecurity media, in various conference sessions, and of course on social media. The good news is that as the awareness of these trouble areas has grown, strategies have arisen for dealing with them.

9.1.1 Burnout

Talk to any career coach or human resources specialist about challenges they are worried about, and *burnout* is likely to be near or at the top of their list. Burnout is a state of exhaustion in one's role. It is usually brought on by prolonged emotional and/or physical stress. A person experiencing burnout will typically feel drained emotionally and may exhibit high levels of anxiety or feel overwhelmed with the demands of their job. Burnout is in no way unique to cybersecurity. It transcends all industries and is always a key business risk for any employer.

In recent years, however, burnout has become an increasingly dominant topic of discussion in cybersecurity circles. Recall at the opening of this book in chapter 1, I talked about the skills shortage that organizations have reported. This was revisited in chapter 3, where I shared statistics showing that much of the shortage is self-inflicted by hiring practices. Regardless of the cause, however, the struggles that organizations have in locating cybersecurity staff put tremendous demand on existing practitioners. These increasing demands have generated higher stress levels across the cybersecurity community.

However, the inability to find skilled resources isn't the only thing putting additional demands on existing staff. In many organizations, cybersecurity is not their core line of business. Therefore, security staff are looked at as an expense or cost-center. They are regarded as a necessary cost of doing business that does not directly generate revenue for the organization. As such, when leadership looks to improve profitability by reducing or eliminating expenses, cybersecurity is often one of the areas of the organization that feels the strain. Despite the growth in awareness and prioritization of cybersecurity in general, overall budgets for security staff and tooling have not grown at a commensurate rate. Staff and capability growth do not ultimately match up to the overall growth of the business. This creates additional strain, but it doesn't stop there.

Much of that business growth is the result of new technologies and innovations. The role of cybersecurity in the face of swiftly developing technologies was discussed way back in chapter 1. As new technologies are launched, cybersecurity professionals are tasked with ensuring the security of those technologies. Within cybersecurity, we are faced with an ever-expanding landscape of systems to defend. We have to constantly be aware of the latest trends and innovations. We must work tirelessly to continue learning and expanding our areas of expertise to ensure we can defend this digital way of life. Add this to the stressors of the staffing and budget situations, and it starts to form what seems like an inescapable loop of trying to do more with less.

Finally, there is the perceived criticality of our role as cybersecurity professionals. After all, it is our job to defend the organization—and if we fail, the whole company could falter. Our customers could have their data exposed, money stolen from them, or other terrible outcomes. The business could lose profits, be fined by regulatory bodies, or lose shareholder confidence. Depending on the industry, whole markets or even worldwide economies could be impacted, as we have seen in recent years with various data breaches and ransomware attacks. This only adds to the level of pressure and stress that cybersecurity professionals experience on a daily basis.

As you enter your cybersecurity career, you will likely experience a *honeymoon phase*. This is that time after taking on a new role when every experience is filled with excitement over the new challenges. You feel energetic as your days are filled with opportunities to learn and grow. Your satisfaction with your job is likely to be its highest at this point.

Unfortunately, that phase does not last forever. As you settle into your role, you will slowly start to experience the stresses that go with it. Day-to-day experiences will become more mundane and repeated. Frustrations will develop over certain tasks or organizational difficulties. This is not unusual and is not unexpected in any career path. Nonetheless, if you do not begin taking steps at this stage to avoid it, stressors will begin to pile up, and burnout will set in.

A key facet in preparing for and preventing burnout is to make self-care a priority and a regular part of your routine. What this self-care looks like can be different for different people. Ultimately, the core needs to be rooted in giving yourself regular and intentional breaks from the stresses of your job. This means avoiding overwork. As a result of the criticality around cybersecurity in the context of the organization and possibly even worldwide markets, overwork is common. You might find yourself feeling a certain level of responsibility for every outcome that drives you to work long hours beyond what is expected of you.

Part of practicing self-care is planning specific times when you will not do work. Turn off the computer, leave the office, and so forth. To help ensure that you prioritize these times, planning other activities can be helpful. Hobbies, physical fitness, family time, or other non-work-related activities can help ensure that you break away.

Using vacation time is another critical element here. It sadly is not uncommon for people to neglect to take the time off that they so diligently negotiated for in their job hunt. In many companies, if you don't use that time off, you lose it. This is time that your employer factors into the expense of employing you, so you should never feel guilty about using it.

When you take time off, be equally intentional about it. If necessary, go to places where you simply cannot be reached, or at minimum, set the boundary before you leave on your vacation that you will not be reachable. Resist that temptation to "just take this one call," no matter how important it seems.

Failing to properly manage burnout by setting boundaries early will not only impact your current job, but can derail your entire career. Professionals often indicate that they almost left or did leave the cybersecurity community simply because they burned out in their jobs. As you sit today, you have a passion and a desire that will be your assets. But you need to maintain that energy to progress along your journey.

9.1.2   Gatekeeping

Inevitably, at some point in your career journey (likely multiple points), you will experience *gatekeeping*. This is the attempt or practice of establishing irrelevant or erroneous requirements for a role or function. For instance, in conversations with other cybersecurity professionals, you may hear that a person must spend time working on a help desk in order to be effective in a security role. But while this role will arm you with certain helpful skills, it is by no means a requirement.

Gatekeeping like this is an unfortunate reality in cybersecurity. Within our community, it seems to be more pronounced than in other industries. It seems to stem, at least in part, from the lack of a clearly defined career progression path. Many opinions exist about what the path should be, and some will take it upon themselves to declare their opinions as facts. Additionally, as discussed in chapter 8, the cybersecurity community is susceptible to toxic levels of competitiveness. This perceived competition leads some to want to invent obstacles that hold others back.

Gatekeeping not only adds to the challenges of impostor syndrome, but also can make people feel like the community is less than friendly overall. It is hard to say how many potential security professionals are turned away because they hear experienced voices saying that they must have some degree of nonsecurity experience before they can enter this field. Anecdotally, these stories are brought to light from time to time, and it can be painful to hear just how unjustified some of these artificial barriers are.

Overcoming gatekeeping is usually not too difficult with just a certain level of awareness. Knowing that this is an activity that some engage in can be enough to tell when their directives are indeed factual versus just their biased opinions. In those cases, simply ignoring their edicts and proceeding on, knowing that many paths could be followed, is often enough.

However, the challenge intensifies when the gatekeeper is in a position to impact your career progression. A person who has a level of influence or authority over you achieving your next goal can be harder to ignore. When you encounter a gatekeeper who is limiting your advancement in some way, the first and most important rule is to keep calm and not attack them. Because of their position of influence or authority, chances are, right or wrong, that they will be believed more readily than you should a confrontation arise.

Keeping your cool does not mean you need to be submissive, however. In fact, it is important to maintain your confidence and show that confidence in a friendly manner through your interactions with that person and others. Be straightforward about your opinions and knowledge and resist the temptation to offer concessions you do not believe in.

Sometimes the easiest way to overcome a gatekeeper is to simply work on making them an ally. Instead of trying to convince them that they are mistaken, simply begin showing them how you can work with them. Assuage some of their insecurities (which are likely at the root of their gatekeeping behaviors) by showing them that you are not a threat. Each gatekeeper you encounter will need to be handled in their own way, but understanding when you are being held back by gatekeeping and having a plan to overcome will help accelerate your career growth.

9.1.3   Stagnation

Believe it or not, despite the demand for constant learning and growth that comes with a career in cybersecurity, many professionals find themselves stagnating after a while. *Stagnation* refers to the idea of falling into a rut, being in a role where you are no longer growing and leveling up your skills or knowledge.

In some cases, stagnation is a result of other personal factors that lead a person to lose motivation for advancing themselves. Some cybersecurity professionals become complacent or comfortable in a given role and stop challenging themselves to do and learn new things. This can be a by-product of burnout, as the loss of passion and love for the job leads to just going through the motions. However, it can also be a result of impostor syndrome or just plain fear of taking a risk. Pursuing a new challenge can be daunting. For some, that risk of trying and failing is just too much to overcome, so they do not even try.

Stagnation can also be a result of the job you are in. Organizations can also become comfortable and complacent with their resources. Poor leadership in particular can lead to a lack of employee development. Additionally, in companies where cybersecurity is not a revenue-generating part of the business, there may be little desire to expand the capabilities and knowledge of that team beyond its current state. The status quo is perceived as enough to keep the business running, and no further investments are made. Finally, depending on the organization, a path to advancement may simply not exist from where you sit currently. Whether it is because vacancies are not opening above you or you have gone as high as you can in that organization, this can be a tough challenge as well.

How you choose to address or avoid stagnation will require you first to identify the cause. Managing your burnout proactively and cultivating your passion and risk taking can help you avoid self-imposed stagnation. Look for projects either within your organization or independently that you can engage in to continue building and expanding your skill set. Look for professional challenges that excite you and shoot your shot. Take the risk knowing that you have plenty of cushion to fall back on if it doesn't work out. After all, this is a high-demand field, and it is not like you are going to end your career by taking a leap of faith.

If you are noticing stagnation because of the limitations of your job, that can be a bit more challenging. Still, look for ways that you can expand your influence across the organization. Can you lead the charge on implementing a new process, a new tool, or something similar that would show you are innovative and forward thinking? You'd get to learn new skills and potentially explore exciting new technology. Perhaps look for ways to engage with higher-level leaders as well. You can talk to your current manager to explore ways to get more visibility within the organization. Finally, consider laying out a personal career development road map. Describe training or other learning opportunities that you'd like to pursue and explore having your employer help fund them.

If these steps fail or your current organization provides no room to advance, you may have to face the reality of a job change. That is OK and is, of course, part of progressing in your career as well. The important facet is to recognize the stagnation early and begin looking for that next opportunity before the stagnation becomes a complicating factor in a job search.

I have maintained the rule that if someone approaches me about a job that seems like it would be a good opportunity for me, I always at least hear them out and explore it. Even if you're not actively looking for a job, that could be one of those serendipitous moments that could lead to your next big thing. There is nothing wrong with exploring new opportunities as they present themselves. Make sure you have a path to keep growing, one way or the other. Figure 9.1 summarizes the three challenges presented in this section as well as tips for overcoming them or avoiding them altogether.

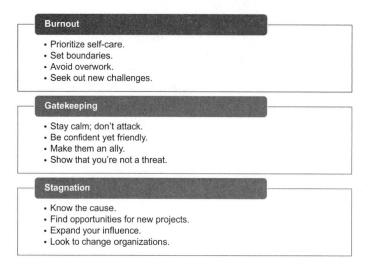

Burnout

- Prioritize self-care.
- Set boundaries.
- Avoid overwork.
- Seek out new challenges.

Gatekeeping

- Stay calm; don't attack.
- Be confident yet friendly.
- Make them an ally.
- Show that you're not a threat.

Stagnation

- Know the cause.
- Find opportunities for new projects.
- Expand your influence.
- Look to change organizations.

Figure 9.1   Common challenges to career success and tips for conquering or avoiding them

9.2   *Building your career strategy*

Building out a career path strategy is an invaluable step that allows you to clearly identify, understand, and achieve the successes you hope for in your professional life. It is not uncommon for people to launch a new career journey without taking the time to truly consider where they are headed. Unfortunately, this can lead to many changes in direction that can slow progression toward an ultimate success goal.

That is not to say that exploring various paths has no value, nor do I mean to suggest that you should shy away from taking advantage of unanticipated opportunities as they present themselves. However, it is good in those moments to have a clear overall understanding of your vision for your career. This will allow you to objectively consider how those opportunities fit into your overall plan. You can then recognize whether you want to adjust your plan or write off these options as simply a distraction that you want to avoid.

The process begins by setting the long-term goals that you would like to achieve, or as I'll reference it, your *vision*. That, as you would expect, is the starting place for any career strategy. However, it is only the starting point. Once you know where you are headed, you need to look at where you are today and identify any existing gaps that will require you to develop new skills or knowledge to achieve that vision. Setting out this list of personal growth needs will help you chart a path to your ultimate goal.

Finally, you need to lay out the path by which you hope to accomplish that vision. Set short-term, achievable, and progressive milestones that will be achievements you can celebrate along the way. This will build your confidence and affirm for you that your career is progressing and not becoming stagnant. That, as discussed in chapter 8, will also help you avoid the damaging effects of impostor syndrome.

9.2.1 Describe your long-term vision

Having a long-term vision of your objective is crucial as you embark on your new career journey. A vision serves as a guide for decision-making all along your career journey. The vision can change over time, but it will be that view of the horizon that you are constantly walking toward. You might even use the cliché of referring to it as your north star. The key is knowing how to set an appropriate vision that is achievable and realistic but also not so narrowly focused that it has to constantly be shifted.

Think about it like navigating a road trip. Before you set out, you likely have an ultimate destination in mind. Before you can decide which highways to take, which scenic routes you may want to explore, you typically first pick out that ultimate destination. Your vision serves as that ultimate destination. It does not mean that once you get there, you are done. Certainly, you may decide to travel farther and visit another destination, but the ultimate destination guides each navigation decision you make along the way until you reach it. Each detour you take needs to be measured with the eventual return to your chosen path. You can explore those unanticipated scenic passes, but ultimately your destination remains unchanged.

Laying out a career vision should be achievable, but not so simple that it can be accomplished too quickly. I generally recommend that people think about where they want to be 7 to 10 years from now. This length of time enables us to generally envision some aspects of the future while at the same time recognizing that the landscape or our goal may change. A longer time frame makes the risk of change both personally and in the industry a near certainty. As a result, maintaining a consistent direction can become problematic. A shorter vision does not allow us to plan a course that results in a high-level achievement and leads to a tactical level of decision-making that lacks an overall clear direction.

Take a look at the self-reflection exercises you did in chapter 4. Consider the areas of cybersecurity that you find most interesting and have identified as your desired career path. Now here comes the hard part. Assuming you start down that path today, where do you see that taking you in 7 or even 10 years? Where do you want to be in 10 years? These sound like nebulous questions, especially since cybersecurity does not always have a good career progression map for us to follow. So, let's talk about how you can figure out a realistic 10-year goal.

The best place to begin is to identify what is most important to you or what indicates success to you. Is a job title something that you feel strongly about and perhaps could serve as that point on the horizon for you? Perhaps salary growth is the thing you are most concerned with. Do you want to progress within a certain industry or perhaps even have goals of working for a particular company or type of organization?

For each person, these factors can be different and carry a different amount of weight. To figure out which factors are most important to you, start by grabbing a piece of paper, and let's do a short exercise:

1 List the various factors that could be measures of success for you. These could include salary, job title, organizational industry, leadership, activism, or community involvement, for example. Be creative.

2 For each factor, go back and try to set a goal for what you would want 10 years from now. Try to be realistic, but if you are unsure of your answers, that is OK at this point. The accuracy or achievability is not as important at this step.

3 Go through each of these factors and goals and rate each individually in order, indicating which ones are most indicative of success to you and which ones you could do without. If you have 10 of them, number them 1 to 10 in order, from most important to least.

4 Beginning with the highest number (the least important), for each one, ask yourself whether you could still feel successful if you did not achieve the goal for that factor but did achieve all the ones above it. If the answer is yes, you would still feel successful, then cross it out. Continue moving up the list in this way until you answer no.

5 The factors that remain in your list are the most important aspects to you for the next 10 years and represent your areas of focus. Now you can use these to craft your vision statement.

Without a clear career progression guide to follow, your network and your mentors can serve as invaluable resources when it comes to finding an achievable objective. If your network includes people in the same discipline that you want to get into, begin reaching out to find out how their careers progressed. Perhaps they can connect you with someone who is at that point in their journey if they cannot help you themselves. Talk to your mentors if you have identified any. Even if they are in a different security field, ask them for ideas or to serve as a sounding board for yours. They should be able to help you validate that your vision is appropriate and achievable in the next decade.

9.2.2 *Identify your personal growth needs*

Now that you have a personal vision statement describing where you want your cybersecurity career to take you, it is time to figure out what you need to get there. The best place to begin is by identifying the knowledge and skills you need to develop as well as the experiences you expect will help you get there. These can be hard questions to answer for someone just getting started in the industry.

So once again, think about how you can leverage your personal and professional networks to gather information. Explore the topic with your mentor and determine whether they can help provide you with guidance or connect you with someone who can. If your goal is to achieve a particular job role, look at job descriptions for that role that you can find online.

Be sure that as you explore what your goal entails that you do not fixate on only technical skills. Think bigger in terms of the leadership, business, and interpersonal skills expected for someone who has successfully achieved that same vision. Think in absolute terms. What are all the facets of that vision that need to be accounted for, whether they are something you currently possess or something you will need to develop?

Once you feel confident that you have a clear understanding of what your vision is going to require, now you can go back to that capabilities inventory that you created in chapter 4. It is time to start looking at the gaps between where you are today and where you want to be in 10 years. The most important part of this exercise is being honest with yourself. Be as objective as you can in terms of where you stand today and how big that gap is. Begin a list that highlights the key areas you need to address in order to be ready for that vision that you have set for yourself.

9.2.3 *Build a 1, 3, 5 plan*

A *1, 3, 5 plan* is commonly used in the corporate world to manage employee performance and development. However, there is no reason you need to wait until your employer requires it to create one. Now that you have a vision and you know the gaps that need to be addressed in order to achieve that vision, a 1, 3, 5, plan will help you lay out the path for getting there, by setting goals for those 1-, 3-, and 5-year time frames.

You may be wondering why this plan extends for only 5 years, when the vision you crafted is for 10 years from now. There are a couple of reasons. First, keeping this detailed plan to 5 years makes it more manageable from a scope perspective. Trying to plan out 10 years' worth of career development can be difficult all at once. The objective here is to make things easier, not to send you down the path of complex planning and futuristic predictions.

Second, limiting your plan to the next 5 years gives you the opportunity to take stock of where you are when you reach the halfway point of your vision. While each of the intervals (1, 3, and 5 years) has its own measurable objectives, as you reach that 5-year objective, you now have the chance to assess whether your plan still has you on target to achieve your longer-term vision. It also gives you enough time to fully understand whether that path to your destination is one that you will really enjoy. Over a 5-year span, you may find that the vision you laid out is no longer the path you want to pursue. So, capping your initial planning at 5 years has less of a lock-in effect than if you laid out the next 10 years of your career in detail.

Building out your 5-year plan should not seem too daunting if you have taken the time to document your vision and list your personal growth needs. This plan simply sets realistic and measurable milestones for addressing those personal growth items. As the name suggests, you will want to lay out the elements that you plan to have achieved by each of the specified time frames.

Making sure that these plans are achievable and realistic is important to the process as well. For this, you will again want to make use of your mentors and network. You may also want to discuss these goals with your current manager or even fellow colleagues in the workplace. Remember, however, that as you gather information from these people, they offer educated opinions, not empirical facts. So take only what you can use and makes sense to you, and let the rest go.

Your 1, 3, 5 plan should be milestones based, as shown in figure 9.2. At minimum, it should include what you plan to have accomplished or where you want to be at each stage. If you want to get more detailed, you can lay out a road map for getting there. A road map would entail the specific tasks that you need to accomplish to reach those milestones. Those tasks could be training, job progressions, or other facets of skills and knowledge development that lead to your milestones.

Does all of this seem too deliberate? Does it seem like a lot of work that most people would never go through? Well, recognize that we all do this just out of human nature. The only difference is we typically are not so deliberate. You probably have ideas in your head of about that CISO job you want, that cybersecurity company you want to start up, or the type of research you want to be doing. The exercise of getting all this information on paper, however, is important to ensuring your success in getting there. Dreams are great to have, but achievable goals and successfully accomplishing those goals are more enjoyable. This is just one method of holding yourself accountable.

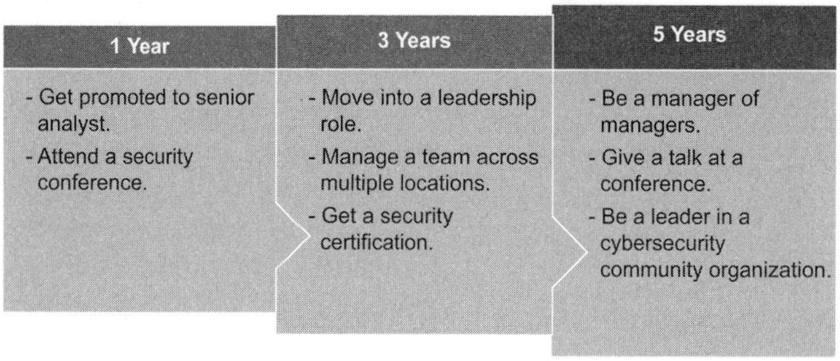

1 Year	3 Years	5 Years
- Get promoted to senior analyst. - Attend a security conference.	- Move into a leadership role. - Manage a team across multiple locations. - Get a security certification.	- Be a manager of managers. - Give a talk at a conference. - Be a leader in a cybersecurity community organization.

Figure 9.2 Example of a 1, 3, 5 plan

9.3 *Pivoting*

One of the truly attractive aspects of growing a career in cybersecurity is the fairly unique ability to easily pivot among domains of specialization. All the various domains discussed in chapter 2 are interconnected in some way. This interconnection is different from many other career paths that have multiple areas of specialty that are not easily transitioned between. For instance, if a neurosurgeon were to decide that she now wants to be an obstetrician, no easy path to do so exists. She'd have to go back through years of schooling and residency to make that change. The tools and techniques are very different and almost completely unrelated in most respects.

Conversely, if a cybersecurity professional currently specializing in security operations wants to move into a security architecture role, much of the knowledge learned in their security operations experience can be leveraged in the architecture space. The challenge with pivoting from one cybersecurity domain to another is not in the execution of the change. Instead, it comes in the form of knowing when it is necessary, charting a path to that pivot, and then taking the risk. The good news is that everything you have done up until now in this guide has given you tools to prepare you to make such a pivot when it is needed or desired.

9.3.1 Recognize when change is needed

While pivoting can be easily accomplished within the cybersecurity domains, it is not to be taken lightly. It still represents a career change that brings with it uncertainty and risk. That said, you might decide to undertake such a change for various reasons. Being able to recognize when the time is right to make a change, sometimes when it is simply needed, can help ensure continued career growth and prevent stagnation.

Usually, the need to make a pivot to a new discipline within cybersecurity will be a personal choice. If you have taken the steps described earlier in this chapter but still find yourself on the brink of burnout, that could be one indicator that pivoting to a new discipline is needed. For example, if you are working in an SOC and the demands and pace of the job are causing too high a level of stress, that might be an indication that a pivot is needed. However, before you jump into the deep end of a career move, it is important to make sure that it is the discipline of cybersecurity that you are not enjoying rather than the way in which your present organization has implemented it.

To this end, it can be helpful to chat with others in your network who are in similar positions. What has their experience been like? Do they share the same frustrations that are making you doubt whether this is the focus area of security that you want to continue working in? If so, that is a sign that a pivot is needed.

You may also want to take a moment to inventory your frustrations. You don't need to write these down if you would rather not, but at least give them some thought. Is it truly the core aspects of your job responsibilities that have you longing for something new? If yes, then maybe a pivot is needed. However, if you're frustrated about the politics of your organization, people you are working with, processes you have to follow, or complications with the tooling or support you are provided, then it may make more sense to explore opportunities in other organizations first.

Stagnation can be another reason for considering a pivot to a new role in cybersecurity. As you progress in your career journey, you may find that you simply do not have a passion for what you are doing. Heading that off quickly and pivoting into a new role could be a way to explore new areas of interest and rekindle the passion you once had. Perhaps over the course of working with other areas of your security team, you have been exposed to other disciplines that seem particularly interesting. There is nothing wrong with deciding to make a pivot simply for the purpose of exploring new domains and building a wider breadth of skill sets and experiences.

Again, the key here is to make sure that a pivot is really what you need in order to address your concerns. If you are keen to explore other aspects of cybersecurity, that can be a valid reason to pivot to a new discipline. If you feel like you have accomplished all you care to accomplish in your current area of expertise, that again is a good reason to look at pivoting. If you are stagnating because no advancement opportunity is available or your employer isn't investing in your personal development, a new role in a new organization might be a better option.

Ultimately, it is important to recognize those career challenges as they come up and how they may lead to a necessity to pivot. You need to be honest with yourself and not be afraid to admit that you need this change. Perhaps you made the wrong choice back when you laid out your interests; that is totally OK. However, also realize that needing to pivot may not be a sign that you made a wrong choice. It can also be a sign that your interests have changed, or you want your career to feature more than that one specialized field of focus. Allow yourself to be OK with making a pivot, and you should never feel like it's a good or bad indication of where your career is headed.

9.3.2 *You need to pivot, now what?*

Remember the exercise for finding your passion that you did back in section 4.2.2? Well, this is a good time to break out those materials once again. Take a look at what you identified as your interests when you made that initial list. Look at how many of those interests you got to explore in your current career path. Also consider how many of those areas of interest still hold true today. This can be a great way of determining whether your passions have changed or you have simply gotten all you can from your initial path and need to change as a result.

If you find that the interests you identified in that inventory no longer apply, now might be a good time to start over and do the exercise again. This will help you identify where you might want to go next. Do you already have an idea of what cybersecurity discipline you want to dive into next? Well, take a look at the passions you identified and see if they are truly a fit. You can continue to leverage that exercise for the rest of your career. Update your list as necessary but also analyze how well your chosen path ends up fitting with your interests.

However you choose the domain you would like to focus on next, this is also a good time to refresh your capabilities inventory. While pivoting within disciplines of cybersecurity is easier than in other careers, you still need to expect to have to learn new skills. In this case, understanding your current capabilities will help you better determine how easily and at what level you can pivot into a new role. If you need to improve or develop skills over those you have today, you now have a nice advantage as compared to when you were trying to land that first job.

You are now working in cybersecurity, and there is a pretty good chance that you can probably start developing those skills in your current role while you look for a position that will let you pivot. Additionally, because you are established within your organization, you might be able to more easily identify and move into a position in that new discipline of cybersecurity without changing companies. Most organizations have a formal process for employees to move between open positions. Take a look at that possibility in addition to looking at external job openings. While the transition timeline may be longer than the notice period you might give if you went elsewhere, the slower process can sometimes be easier and allow you to better prepare for success in your new role.

Revisiting a topic from chapter 6, be careful to research and know your worth in the new position. One challenge when making a pivot within the same organization is that the company may be less willing to significantly increase your salary even if the new position warrants it. Do the research necessary to know what you should be making in that new role, whether it is an internal transfer or a move to a new company. Pivoting can sometimes mean moving into a role that is more highly compensated, but that is not always the case.

Looking at your capabilities inventory, you may find that while you are in a senior role today, you are qualified for only a more junior role in the new domain you are pivoting to. In this case, you need to determine whether a salary reduction will be necessary and whether that is acceptable to you. Again, this is an opportunity for you to research and know the worth of that new position before you go into the interview and negotiation process. Ultimately, the biggest challenges in making a pivot are less about the skills required and more about the logistical aspects of the new role.

9.3.3 *Shoot your shot*

Risk taking is a topic that has come up quite a bit in these last two chapters. When deciding to pivot in your career, you are taking a risk. You are leaving the familiar, whether it has been a positive or a negative experience, and you are moving into unknown territory. The good news is that this is a high-demand field, and you could always pivot again or even pivot back if you feel like that is right for you.

Taking risks means being aggressive to a degree. If you are going to take on the risk of a pivot, make sure that sufficient potential reward can result. Maybe all you are looking for are new ways to learn and grow. Maybe you have decided that you want a discipline with greater salary potential. Perhaps your pivot is more about finding a way into higher levels of leadership. You need to know your motivations and weigh the potential rewards in terms of what is most important to you.

When it comes right down to it, taking that leap is a necessary part of growth. As discussed in chapter 8, if you become complacent and do not take risks, you will likely stagnate, which can have long-term implications on your career. When you've decided that it is time for that change of scenery and you want to pursue a new avenue of cybersecurity, don't let impostor syndrome or your need to level up certain skills stand in your way. You did not let them stop you from going for that first job. You didn't let them get in the way of setting ambitious goals for yourself, so why would you let them stand in the way now as you look to continue growing?

9.4 *Putting it all into motion*

You've made it! In chapter 1, you learned about what cybersecurity is and how it has become a crucial and inherent part of our everyday digital way of life. You discovered that diversity is a necessary component to the success of our endeavors. In chapter 2, you learned about the many exciting and varied roles that exist within cybersecurity. You saw how each fits into an overall objective and plays a crucial role in defending digital systems. You also learned more about the traits that make for the best cybersecurity professionals and learned which practices are best avoided.

In chapter 3, we started taking a look at job progression and technical skills that are often called upon in cybersecurity roles. You also saw how soft skills can be equally if not more important than technical skills. In chapter 4, you learned about the challenges that you will likely encounter when trying to land your first cybersecurity job. You went through multiple self-analysis exercises meant to help you find your desired career path and prepare for the job hunt. You discovered what core skills are and how they can help you demonstrate your value for a cybersecurity role in lieu of other desired technical skills and experience.

Chapter 5 explored certifications and other ways to build technical skills in cybersecurity. Chapter 6 provided strategies for mastering your resume and being successful in the interview process. Chapter 7 explored building your personal network and establishing mentorship relationships.

Chapters 8 and 9 changed the focus to the future, equipping you with tools that will help ensure your long-term professional success in cybersecurity. And now, here you are, ready to become a formidable force in the fight to protect our digital way of life. You are ready to take on the challenges of being a cyber defender. You are positioned to enjoy all the exciting and lucrative aspects of working in a career field that is in high demand and promises to be for years and decades to come.

What one last piece of advice can I offer to you that I haven't already covered in the hundreds of pages of this guide? Let me leave you with this: The cybersecurity community is an amazing collective of highly driven and highly skilled people. The earliest days of hacker culture still influence the ways that we interact even today. Take pride in becoming a part of this wonderful community, and with that pride I hope you will also help shoulder the responsibility of making this community even stronger. That means actively working to improve our inclusivity, focusing on what matters most, and elevating those who come next.

By uplifting one another, we make our entire community better and our world just a little bit safer tomorrow than it was yesterday. I wish you all the greatest of successes in your career journey and I hope one day you and I can talk about your experiences.

Summary

- Cybersecurity professionals have to deal with unique challenges in the form of burnout, gatekeeping, and stagnation.
- Setting long-term goals and building a progressive strategy to achieve them can help avoid those challenges and conquer impostor syndrome.
- Pivoting from one area of cybersecurity to another is an expected outcome and can be more easily accomplished in this field than other industries.

glossary

While the following is not an exhaustive list, it does cover many of the common cybersecurity terms that are introduced throughout this book or that you may encounter as you explore launching your career. The intention is to give you some exposure to these terms so you can conduct your own additional learning to fully understand them.

advanced persistent threat (APT)   Specific threat actors or threat actor groups who stealthily gain access to a system and maintain that access for an extended period of time to compromise additional systems and expand their access and impact.

allow list   A list of known values that are permitted to pass a particular security control, also known as a *white list*, although that term has fallen out of favor.

application security   The collection of practices within an organization designed to ensure the security of software developed by the software engineering teams.

ARPANET   The Advanced Research Projects Agency Network, an experimental network of networks that connected the computer networks of various independent entities together and was the predecessor of what we know today as the internet.

asset inventory   A catalog of all known digital and physical information technology within an organization.

authentication   The process of determining whether an entity is who they claim to be.

authorization   The process of defining and/or verifying that an entity is allowed to access a particular resource.

blue team   Cybersecurity professionals who are engaged in defending digital systems and users from attackers.

botnet   A group of computers that have been compromised by an attacker for purposes of launching additional attacks.

breach   An event in which an attacker is able to successfully bypass security controls to gain unauthorized access.

buffer overflow   A type of attack in which the attacker is able to overwrite system memory and thus modify data or the instructions being executed by the computer system's processor.

bulletin board system (BBS)   Computer software that allows users to connect, usually via a modem, and interact with other users in a text-only environment. Users can upload/download files, read posts from other users, and post their own information. The BBS was a popular way for hackers to interact and share information before the age of the internet.

business information security officer (BISO)   The leader of a cybersecurity program for a division, group, or business line within an organization.

call for papers (CFP)   A solicitation of members of the community to propose presentations and other educational sessions at a conference or event.

capture the flag (CTF)   An activity in which users are tasked with finding various indicators (flags) within a system, typically by executing some form of attack. Often these are set up as competitions, with hackers attempting to break into various aspects of a system to find the flags.

chief information security officer (CISO)   The executive leader of an entire organization's cybersecurity program.

ciphertext   Data that is created by passing known data through an encryption process.

cloud   A service providing computing resources that people and organizations can use, typically in lieu of building and maintaining their own information technology systems.

command and control (C2)   The centralized system or network that controls the members of a botnet from which attackers can launch their attacks.

Compatible Time Sharing System (CTSS)   A computer system launched by MIT in the early 1960s that allowed multiple users to access the same system simultaneously. In addition to being the first system to do so, it is credited as the first known computer system to leverage user passwords to authenticate multiple users.

compromise   Exposure of a resource to an unauthorized attacker.

Computer Emergency Response Team (CERT)   A division within the Software Engineering Institute of Carnegie Mellon University that studies cybersecurity issues and works with various entities in the government and industry to create solutions.

confidentiality, integrity, availability (CIA) triad   A model for discussing the goals of security controls. *Confidentiality* refers to protecting resources from being viewed by unauthorized parties. *Integrity* refers to protecting resources from unauthorized modifications. *Availability* refers to keeping a resource available for access by authorized parties.

configuration management database (CMDB) A catalog of information about the information technology assets within an organization and detailed descriptions of how they're configured.

control A safeguard or countermeasure that is put in place to reduce the risk of system compromise.

cross-site scripting (XSS) A form of web application attack that allows the attacker to cause unintended data to be returned to a victim's web browser. This attack is described in more detail in the OWASP Top 10 Web Application Security Risks.

cryptography The practice of protecting data from being accessed by unauthorized parties by applying a complex mathematical rule to the data to change it in such a way that it cannot be reverted without knowledge of the rule used.

Cybersecurity and Infrastructure Security Agency (CISA) A US government agency, part of the Department of Homeland Security, created in 2018 to help manage cybersecurity across government agencies and the nation's critical infrastructure.

cybersecurity industry A community of people and organizations interested in protecting digital systems throughout all facets of our society.

data center A physical facility that houses an organization's information technology systems and provides the required power, air cooling, and other infrastructure for their continued operation and use.

data loss prevention (DLP) The practices and controls implemented by an organization to protect against users exposing confidential information to unauthorized parties via intentional or unintentional means.

denial of service (DoS) A form of attack in which the attacker seeks to make a particular system or resource unavailable for use.

deny list A list of known values that are not permitted to pass a particular security control, also known as a *black list*, although that term has fallen out of favor.

DevOps A model of software development in which software engineers (developers) work together with the teams that support the software once it is complete (operations) to make the process as efficient as possible.

DevSecOps The integration of security practices into the DevOps model.

digital certificate An electronic key, also called a *public key*, used to encrypt data before it is transmitted or stored in a location. A private key is then required to reverse the cryptography to decrypt the data.

digital forensics A cybersecurity discipline that focuses on analyzing various aspects of a system to determine events that have occurred and potentially preserve evidence of those events.

digital forensics and incident response (DFIR) The combination of two related disciplines of digital forensics and incident response.

disaster recovery (DR)   The process by which an organization is able to respond to events that adversely impact their systems or their ability to conduct business and restore services.

distributed denial of service (DDoS)   An attack in which the attacker uses a large number of systems, often from a botnet, to launch a DoS attack against a victim's system. The aim is ultimately to overwhelm the target system and cause it to be unavailable for legitimate use.

egress   The process of data leaving a system.

encryption   Applying a cryptographic method to data to make it unreadable to unauthorized parties.

ethical hacker   A hacker, usually employed or contracted by an organization, who attacks the organization's systems in an attempt to identify security vulnerabilities and determine how they can be fixed. Also referred to as a *penetration tester.*

exploit   The act of using a security vulnerability to gain unauthorized access to a system. The term is also often used to describe the actual tactics involved.

firewall   A network device or software that controls access to a network or the resources within the network by analyzing network requests and applying rules that specify what should be allowed or not allowed.

fuzzing   An automated technique for testing a system for security flaws or coding errors by sending various forms of invalid or unexpected data.

governance, risk, compliance (GRC)   A strategy for managing various aspects of an organization's information technology approach to ensure that it supports the business. *Governance* refers to using policies and standards that ensure that processes support the business goals. *Risk* refers to identifying and responding to factors that may negatively impact the accomplishment of business goals. *Compliance* refers to ensuring that the business's practices meet requirements of laws and regulations that apply to the business.

handles   The pseudonyms people use as identifiers on various social media or other platforms. Handles are often leveraged as a way to maintain a level of anonymity.

hash   A cryptographic method producing a string of characters that is of a fixed expected length and cannot be reversed back to the original data.

honey pot   A decoy system designed to trick attackers into launching their attacks against it, allowing defenders time to detect and respond to the attacker before they target legitimate systems.

identity and access management (IAM)   A framework of practices, processes, and technologies for providing authentication and authorization to information technology systems.

incident response (IR)   The process or discipline for responding to events that could adversely impact (or already have impacted) information technology systems and/or the business processes they support.

indications of compromise (IOC)   Data found through digital forensics techniques that potentially identify malicious activity on a given system.

industrial control systems (ICSs)   Information technology systems that are used to manage physical systems such as manufacturing machines, utilities instruments and controls, and so forth.

information security   The function of a business associated with protecting its information technology assets from threats.

information technology   The set of digital systems (including computers, networks, and peripheral devices) that organizations rely on to conduct business.

ingress   The process of data entering a system.

insider threat   People within an organization who could potentially cause harm to the organization by intentionally or unintentionally compromising systems or data.

Internet of Things (IoT)   Devices people traditionally use in everyday life whose purpose is not to provide computing capability yet contain some level of computer processing to allow them to connect to networks and interact with other digital systems (for example, smart refrigerators and fitness trackers).

intrusion detection system (IDS)   A system that monitors the activity on a network or system to identify potential attacks and provide alerts to defenders.

intrusion prevention system (IPS)   A system that monitors the activity on a network or system to identify potential attacks and prevent them from completing successfully while also providing alerts to defenders.

least privilege   A framework for ensuring that users are authorized to access only the minimum functions and resources they require for a particular purpose, task, or job.

malware   Malicious software designed to compromise a computer system, providing unauthorized access to an attacker.

mitigating control   A security approach that doesn't specifically resolve a particular threat but lessens the overall risk posed by that threat.

multifactor authentication (MFA)   The use of multiple forms (factors) of proof to authenticate a user. For instance, something they know (a password) and something they have (a code from a phone app) would be two factors for authentication.

National Institute of Standards and Technology (NIST)   A nonregulatory agency of the US government, founded in 1901 as part of the Department of Commerce, that sets various standards—including for cybersecurity.

Open Systems Interconnection (OSI) model   A seven-layer model describing the various functions that allow computers to communicate over a network.

Open Web Application Security Project (OWASP)   A nonprofit organization, founded in 2001, that focuses on various projects and initiatives to improve the security of software.

penetration tester   A hacker, usually employed or contracted by an organization, who attacks the organization's systems in an attempt to identify security vulnerabilities and determine how they can be fixed. Also referred to as an *ethical hacker*.

phishing   Sending specially crafted emails in an attempt to get the recipient to respond in a way that exposes their private data or allows malware/ransomware to compromise their system.

piggybacking   Knowingly allowing a second person to pass through a control point (such as a locked door or turnstile) without authenticating themselves first.

port scan   The act of attempting to determine what services a network-connected system is running by sending nonmalicious traffic to the system to determine how it responds.

purple team   A team comprising members of a red team (conducting attacks against a network or system) who work in collaboration with the defenders (blue team) to improve defensive controls to block those types of attacks that were successful.

ransomware   A specific type of malware that, after compromising a system, encrypts files and data to make them unavailable to the user or organization and then demands a ransom to be paid in order to decrypt the data and recover access to it.

red team   Cybersecurity professionals who are engaged in identifying security vulnerabilities in systems and software by attempting to mimic the tactics and techniques of attackers.

remote code execution (RCE)   The result of exploiting a security vulnerability in a way that allows the attacker to run unauthorized commands on the compromised system.

repudiation/nonrepudiation   The ability to deny the validity of a piece of data (repudiation) or the assurance that such disputes cannot be made (nonrepudiation). For instance, providing indisputable identification of the user responsible for executing a specific action on a computer system would be non-repudiation.

reverse engineer   To deconstruct a piece of software down to some level of source code representation so that its functionality can be analyzed without executing it.

risk   The probability that a specific threat will be able to compromise a system. Typically, it is a value derived by considering the likelihood of a compromise and the potential impact to the system or business should a compromise occur.

security assessment and verification   The set of processes for analyzing the strength and resilience of systems and software.

security incident and event management (SIEM)   A system that collects information from various information technology systems and provides the capability to analyze and respond to events, often in an automated fashion.

security operations center (SOC) A centralized part of an organization tasked with monitoring and managing security controls that are running in the organization's information technology systems. Can also be used to refer to the specific location or locations where the security operations team is located.

security operations team The team tasked with monitoring and managing security controls that are running in an organization's information technology systems.

shared responsibility The idea that two groups have some level of responsibility for the success of a business objective. Within DevSecOps, this refers to all three disciplines being expected to support efficient development of stable software that is secure. The *cloud shared responsibility model* refers to a division of responsibilities between the cloud vendor and the customer to secure the systems that are deployed in the cloud.

social engineering The discipline of using deception to manipulate the way people respond in a given situation in order to bypass security controls.

software development life cycle (SDLC) The set of repeatable processes by which an organization's software engineering teams create, test, and deploy software.

software security The collection of practices within an organization designed to ensure the security of all software that is deployed in the organization's environment, whether created by the organization or provided by a third-party vendor.

spear phishing Sending specially crafted emails in an attempt to get the recipient to respond in a way that exposes their private data or allows malware/ransomware to compromise their system. *Spear* refers to using specific knowledge of the targeted person to create a more convincing false email.

spoofing Disguising the source of a particular request or action in an attempt to bypass security controls or obscure the entity responsible.

SQL injection (SQLi) A type of application attack in which an attacker can execute commands against an application's database by sending specially crafted data to the application's user interface.

tactics, techniques, and procedures (TTP) The set of common activities that make up a pattern of attack and that can then be attributed to a particular type of attack or even group of malicious attackers.

tailgating The process of a second person passing through a control point (such as a locked door or turnstile) without the knowledge of the authenticated person ahead of them.

threat A malicious action or actor seeking to compromise systems for various purposes.

threat modeling Analyzing a system to understand the likely threats it faces, with the goal of proactively designing countermeasures to address those threats.

virus A subset of malware, this malicious code is designed to impact a system's functionality or gain unauthorized access to the system while also spreading itself to other systems.

vulnerability   A flaw or weakness in a system that could allow an attacker to bypass other security controls of the system.

vulnerability management   The practice within an organization of actively attempting to identify and remediate security vulnerabilities across systems and software.

web application firewall (WAF)   A specific form of firewall that analyzes requests sent to a web application to identify and defend against potential attacks.

worm   A subset of malware, this form of malicious software spreads itself from one system to another via the network. It is different from a virus in that it doesn't necessarily impact system functionality or gain additional access.

zero trust   A model for implementing security controls in which all components of a system are treated as untrusted by all other components of the system, and therefore all interactions must first go through authentication and authorization.

index

RELATED MANNING TITLES

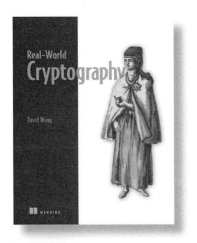

Real World Cryptography
by David Wong

ISBN 9781617296710
400 pages, $59.99
Publication in September 2021

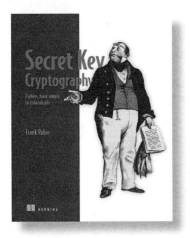

Secret Key Cryptography
by Frank Rubin

ISBN 9781633439795
225 pages (estimated), $49.99
Publication in June 2022 (estimated)

For ordering information go to www.manning.com

RELATED MANNING TITLES

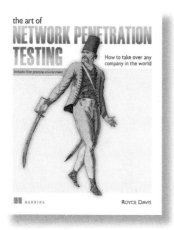

Art of Network Penetration Testing
by Royce Davis

ISBN 9781617296826
304 pages, $49.99
Published November 2020

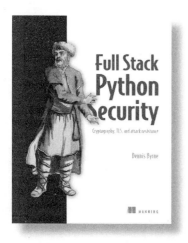

Full Stack Python Security
by Dennis Byrne

ISBN 9781617298820
304 pages, $59.99
Published July 2021

For ordering information go to www.manning.com